# Guided Wave Optics

# Guided Wave Optics

## Alan Rolf Mickelson

*Electrical and Computer Engineering*
*University of Colorado at Boulder*

VNR  VAN NOSTRAND REINHOLD
New York

Copyright © 1993 by Van Nostrand Reinhold

Library of Congress Catalog Card Number 92-30336
ISBN 0-442-00715-9

I(T)P  Van Nostrand Reinhold is a division of International Thomson
Publishing. ITP logo is a trademark under license.

Printed in the United States of America

Van Nostrand Reinhold
115 Fifth Avenue
New York, NY 10003

International Thomson Publishing
Berkshire House
168-173 High Holborn
London WC1V7AA, England

Thomas Nelson Australia
102 Dodds Street
South Melbourne 3205 Victoria, Australia

Nelson Canada
1120 Birchmount Road
Scarborough, Ontario M1K 5G4, Canada

16  15  14  13  12  11  10  9  8  7  6  5  4  3  2  1

**Library of Congress Cataloging-in-Publication Data**

Mickelson, Alan Rolf, 1950–
    Guided wave optics/Alan Rolf Mickelson.
        p.  cm.
    Includes bibliographical references and index.
    ISBN 0-442-00715-9
    1. Integrated optics.   2. Optical wave guides.   3. Lasers.
    I. Title.
    TA1660.M52 1992
    621.36'93—dc20                                    92-30336
                                                      CIP

*To my father, who constantly tried to impress on me that it was the rewriting and not the writing that would make a book good. I am sure he would think this book needed a half dozen more rewrites.*

# Contents

Preface     ix

Acknowledgments     xi

**1 Overview of Guided Wave Optics**     **1**

  1.1 Some Common Guided Wave Systems     1

  1.2 Guidance in Materials     3

  1.3 Components of Guided Wave Systems     6

  References     7

**2 Slab Waveguides**     **8**

  2.1 Geometrical Optics of Slab Waveguides     8

  2.2 Electromagnetic Theory of Slab Waveguides     12

  2.3 Extensions to Guides with Loss or Gain     31

  References     41

  Problems     42

**3 The Semiclassical Laser Equations**     **48**

  3.1 The Derivation     48

  3.2 Rate Equations     71

  3.3 Subthreshold Behavior     77

  Reference     83

  Problems     83

**4 Semiconductor Lasers**     **94**

  4.1 Effective Indices     95

  4.2 Band Structure     112

  4.3 Fermi Levels     117

  4.4 Semiconductor Rate Equations     128

  4.5 Semiconductor Laser Structures     135

  References     137

  Problems     138

**5 Optical Fibers**     **145**

  5.1 Some Fiber History     145

  5.2 The Modes of Optical Fibers     146

5.3  Some Comments on Fiber Fabrication                      165
5.4  Coupled Polarization in Single Mode Fibers              168
5.5  Propagation in Fibers in General                        172
5.6  Dispersion                                              194
     References                                              203
     Problems                                                204

**6 Integrated Optics**                                      **215**
6.1  The What and Why of Integrated Optics                   215
6.2  The Electrooptic Effect                                 221
6.3  Electrooptic Devices                                    224
     Problems                                                241

     List of Symbols Used                                    247
     Index                                                   249

# Preface

This book has grown out of a need for a beginning graduate level text which emphasizes the unifying concepts of the field of guided wave optics. Over the past twenty years, progress in this field has been so rapid, and therefore so helter-skelter, that it is hard, even for the fully involved practitioner no less the aspiring student, to see the unifying concepts. As will be discussed more fully below, there are at present quite a number of texts in the guided wave area. These texts vary in nature from the popular treatise to the voluminous scholarly work. I know of none, however, that treats the waveguide, semiconductor laser, fiber and fiber component, and integrated optic component all on equal footing using the forms of Maxwell's equations in polarizable media and coupled forms of Maxwell's equations as unifying tools. This book emphasizes basic concepts, yet is quantitative in nature and contains numerous applications.

The book is designed to be used by the beginning graduate student or the professional who needs to review or catch up on the basics. Here at the University of Colorado, this text is generally used for a follow-up course to one in Physical Optics. The Physical Optics text employed, also written by the present author, primarily includes material which should be familiar to students with a strong background in optics or practitioners of guided waves. Therefore, although Physical Optics is a recommended prerequisite for the Guided Wave Optics course, it is not in all cases necessary. Here at the University of Colorado, this course is a first-year graduate course taken by students from the Electrical and Computer Engineering and Physics Departments.

This book tries to take a slightly nontraditional approach. Rather than attempting to concentrate on details of operation and therefore emphasizing differences between things, it strives to emphasize similarities between things. For example, the semiconductor laser is presented simply as a waveguide in a material whose index of refraction is dynamical at the frequency of operation. A fiber is also presented as a waveguide, but one which is usually operated in a frequency regime where its loss is low enough that a pump is unnecessary. On another issue, the book attempts to clarify that the description of modal propagation in two mode fibers is equivalent to the description of polarization propagation in single mode fibers is equivalent to the description of energy level coupling in a two-level system is equivalent to the description of the action of an integrated optical directional coupler, etc. The book, in general, tries to tie together as many concepts as can be tied together.

# Acknowledgments

I should acknowledge all the students who have taken this course over the last six years for their contributions and comments which have been a guide in the numerous revisions of this manuscript. I should further acknowledge many useful discussions with my colleagues, especially with Ed Kuester and Dag Hjelme here in Boulder, as well as Audun Weierholt at SINTEF DELAB in Trondheim. Perhaps most important, I should acknowledge a number of my present and former graduate students who have materially contributed to the manuscript, including Walter Charczenko, Chow Quon Chon, Richard Fox, Indra Januar, Holger Klotz, Paul Matthews, Marc Surette, Sandeep Vohra, Mike Yadlowsky, Shao Yang, Peter Weitzman and probably a host of others I have failed to mention.

# 1

# Overview of Guided Wave Optics

This introductory chapter should serve both to preview the contents of this book and to give some philosophical justification for the inclusions in and exclusions from the development.

## 1.1 SOME COMMON GUIDED WAVE SYSTEMS

The thought of a guided wave system conjures up different pictures in different people's minds. In this section, we wish to discuss several of the most commonly used guided wave systems and explain why they are not going to be discussed further in this book.

Perhaps the most common guided wave systems are of the type depicted in Figure 1.1 — that is, imaging systems. These systems, apart from diffraction losses, clearly guide optical energy from an initial lens to an imaging plane. Such imaging systems include cameras, microscopes, and telescopes. In the past, there has also been discussion of lens waveguide systems like the one depicted in Figure 1.2 (Marcuse 1982, Chapter 4). Interest in such systems arose not only from the idea that it might be possible to realize a long distance communication system using gas lenses (Marcuse 1982, Chapter 4) but also from the close analogy between such a system and an "unfolded" laser cavity (Yariv 1985, Chapter 2).

The purported purpose of lens waveguide systems as well as imaging systems is the transmission of information along a preselected path. A system type which shares methodology but not purpose with imaging systems is the directed energy system. An example of such a system is illustrated in Figure 1.3. It is purported that, (see, for example, Clause 1973 or Stavroudis

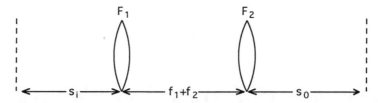

**FIGURE 1.1**    Simple two lens imaging system.

1973) in one encounter during the third century B.C., a hopelessly outclassed Syracuse turned to state-of-the-art technology devised by Archimedes to thwart a coordinated Roman sea invasion. Huge focusing shields were used to focus spots of concentrated sunlight on the bows and sails of invading Roman ships in the hope of repelling the enemy without having to resort to hand-to-hand combat. Such systems achieved a theoretical renaissance in the 1950s with Buck Roger movies and seemed to still be in favor with a selected few in the 1980s and even into the 1990s.

Systems which rely on free space propagation for connectivity tend to be inexpensive and easily constructed, but are not up to the stringent requirements inherent in such applications as optical communications or reiterative optical processing. A free space connected system will always require lumped elements (lenses, beam splitters, etc.) to perform focusing and routing. In a lumped electrical circuit, lumped elements generally cause both series and shunt conductances which lead to lumped losses. This situation holds true also in the optical case, assuming that free space remains free enough (no fog, smoke, rain, etc.) that the lumped losses swamp the distributed propagation losses. Losses are very problematic indeed, as they place sharp limitations on the total propagation distance within a system. Lumped electrical or electronic systems circumvent this limitation by electrical repeating. In integrated circuits, this repeating is performed essentially at every stage. Optical repeating is not performed as easily. Electrical propagation is not associated with diffraction loss, and therefore wires can be thin, even thinner than the "gain" areas that lead to regeneration. In the optical case, one wishes to have a wide enough beam to minimize diffraction. Operations

**FIGURE 1.2**    Schematic depiction of a lens waveguide system.

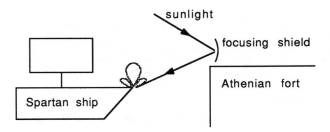

**FIGURE 1.3**    Application of a directed energy system.

such as amplification or detection require small areas (in terms of wavelengths) in order to be fast and efficient, however. As major advantages of optics are low propagation loss and fast switching time, the conflicting requirements of large beam cross section and small lumped amplifier size do not bode well for the application of free space optics to information transmission (with the exception of satellite to satellite communications) and repetitive processing. There are those who would contend this point however.

It is just such problems that the "new" guided wave optics addresses. As operation frequencies moved from radio frequency (RF) to microwave frequency in electronic ciruits, lumped circuits gave way to distributed circuits. A somewhat analogous situation has arisen in optics where imaging systems have given way to guiding structures which have distributed their losses along their lengths. As the wave is guided, diffraction losses are of consequence only at junctions where one guide must couple to another. The early multimode fiber systems were mainly of efficacy in that their losses were lower than those of a corresponding lens guide, they were more rugged than a multikilometer imaging system could be made to be, and the fibers could be continuously bent around corners. Repeating still required complete electrical decoding and regeneration of the optical signal. Single-mode systems of today have additional advantages in that the fiber cross section is well matched to that of integrated optic components as well as to laser sources and amplifiers. It is easy to see that a time approaches when fibers can be used as wires to interconnect optical circuits. As will be outlined in the remainder of this chapter, it is the fiber and the components of these optical circuits which will be the subject of the course.

## 1.2 GUIDANCE IN MATERIALS

Apparently, it was known to the ancient Greeks that a sheet of glass could serve as an integrated optic component, as is illustrated in Figure 1.4. In the

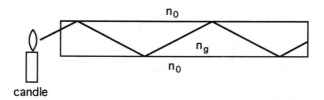

$n_0$

$n_g$

$n_0$

candle

**FIGURE 1.4**    Guidance in a glass slab excited by a candle.

experiment performed at that time, it was noticed that, if one were to view the flame of a candle lengthwise through a piece of glass, one could in general see a (very) distorted image of the candle. As is illustrated in Figure 1.4, the reason for this observation is closely related to the concept of total internal reflection—that is, the property that rays emitted by the source will repeatedly be reflected off the glass/air interface and back toward the center of the glass slab.

Not long after Maxwell's equations became accepted, it was noted that such slab problems as that illustrated in Figure 1.4 could be solved as electromagnetic boundary value problems. A solution of such a problem is depicted in Figure 1.5. The boundary value solution serves as a tool with which to analyze the properties of wave propagating in a structure with attention to such characteristics as cutoff conditions, waveguide dispersion, allowed ray paths, etc. Waveguide boundary value problems and their complex variable counterparts will be the subject matter of the second chapter. A major problem with the simplest form of electromagnetic analysis, in which one assumes all of the refractive indices are real, is that although it could be used to predict coupling losses it cannot be used to predict propagation losses.

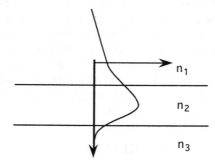

$n_1$

$n_2$

$n_3$

**FIGURE 1.5**    Asymmetric slab waveguide with the electric field structure in the three regions superimposed over it.

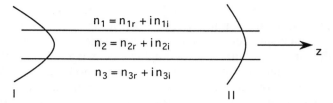

**FIGURE 1.6**    Effect of propagation in a lossy medium ($n_{2i} > 0$) where a forward propagating wave at plane I is attenuated before arriving at plane II.

Studies of guiding media with complex indices are of much later vintage than the earlier loss-free studies. The basic problem is illustrated in Figure 1.6, where now each of the regions will, in general, have a complex index, with the imaginery part denoting gain (if negative) or loss (if positive). Inclusion of these loss (gain) effects extends one's knowledge of the propagation problem considerably. As the equations describing the propagation of the wavefront are no longer real in this case, it is clear that one can no longer find a real propagation constant and, indeed, the imaginary part of the propagation constant gives one the propagation loss which was lacking in the case of real indices. In addition, the phase fronts of a wave with complex propagation constant can no longer be plane propagating, and the complex problem can predict this effect along with the propagation constant.

Although the complex eigenvalue problem is illustrated in Figure 1.6 can give one information about mode structures, it has little or no dynamical content. Yet, for example, in a gain medium the energy associated with the medium is being transferred to the propagating electromagnetic disturbance and, unless it is replaced, will rapidly be used up. To have a steady state gain, one must continuously replenish the "inversion" of the material through a

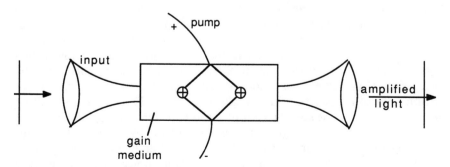

**FIGURE 1.7**    Schematic depiction of a laser amplifier.

pumping process. As the actual amplification (or damping) process of optical waves corresponds to atomic processes, the pump process must somehow parcel the replenishing energy to the individual atomic sites within the lattice. As illustrated in Figure 1.7, since light velocity is a limiting velocity, the dynamics of the replenishment process can occur no faster than the optical readout of this energy. In this case, the index of the medium varies dynamically and its dynamic variation is coupled to both the pump and to the dynamical variation of the light intensity, which in turn is coupled to the index through the wave equation. A careful study of this process, culminating in the semiclassical laser equations for two-level systems, is the subject matter of Chapter 3.

## 1.3 COMPONENTS OF GUIDED WAVE SYSTEMS

Although gas lasers are easily modeled as two-level systems, such lasers are not even candidates for use in guided wave systems. Semiconductor lasers and devices are the active elements of guided wave systems, It is the purpose of Chapter 4 to apply the semiclassical laser equations to the light-semiconductor material interaction. This exposition will begin with a study of the mode structure of standard semiconductor waveguides. This discussion will introduce the effective index method of analysis, the most powerful method for understanding integrated optical devices. A short review of the basic elements of semiconductor physics will be used there to illustrate the relationship between semiconductor atomic dynamics and those of two-level atomic systems. Following this discussion, the semiclassical semiconductor laser equations will be used to derive some of the salient features of semiconductor lasers.

No discussion of guided wave optics could be complete without a discussion of the "wire" which ties the guided wave components together. Indeed, the subject matter of Chapter 5 is the optical fiber. The discussion will begin with some historical background and presentation of an "exact" solution to the problem of the dielectric cylinder. The discussion will continue by showing why the "LP" approximation to the exact solution may actually be a more exact solution to the fiber problem than the "exact" solution. The discussion will proceed by considering the simplest example of fiber propagation, that of propagation in a single mode fiber. This discussion will by nature need to use the so-called coupled mode theory, which will be seen to be nothing more than a slight generalization of the two-level system arguments given in Chapter 3. Certainly, for high bandwidth, long distance applications, the single mode fiber is a superior choice to the larger core multimode fiber. However, as more and more applications for fiber systems arise, more and more applications will appear in which the low cost and low tolerance installation requirements of multimode fibers are the prime system needs. Even if this were not the case, multimode fiber deserves attention if only for exposing the relation between multimode fiber propagation and partial coherence theory. The discussion of the multimode fiber will begin with a description of the disastrous effect of modal noise, an effect which occurs whenever one uses a highly coherent source with a multimode fiber. The discussion will continue with an exposition of the elements of the radiometric treatment of the multimode fiber, the application of this treatment to multimode systems excited by poorly coherent (i.e., light emitting diode (LED)) sources.

The final chapter of the book will center its attention on slightly more futuristic devices—that is, the so-called integrated optic devices. The exposition there will begin with the coupled mode analysis of the archetypical integrated optic device, the directional coupler. Discussion then will go over to realizations of such a device in both $LiNbO_3$ and GaAIAs. This discussion will include an explanation of the fabrication technology necessary. The exposition continues with a description of the operating principles of various other electrooptic components such as phase shifters and Mach-Zender interferometers.

### References

Claus, A. C., 1973. On Archimedes' Burning Glass, Appl Opt *12*, A14.

Marcuse, D., 1982. *Light Transmission Optics,* Second Edition. Van Nostrand Reinhold, New York, See Chapter 4.

Stavroudis, O. N., 1973, Comments on: On Archimedes' Burning Glass, Appl Opt *12*, A16.

Yariv, A., 1985. *Optical Electronics,* Third Edition. Holt, Rinehart, and Winston, New York, See Chapter 2.

# 2

---

# Slab Waveguides

This chapter will concentrate on analyzing one-dimensional guiding media — that is, media in which the index of refraction varies in only one direction and is of such a form that it can trap electromagnetic energy propagating perpendicularly to its index gradient, in the sense that the majority of the wave's energy will propagate in the interior of the guiding region rather than dispersing away with ever increasing coordinate. The standard geometry that will be used for such a medium is illustrated in Figure 2.1. That $z$ is the direction of propagation of the optical wave is a convention that will be used throughout this book. In later chapters, however, refractive indices will be allowed to vary in both the $x$ and $y$ directions.

## 2.1 GEOMETRICAL OPTICS OF SLAB WAVEGUIDES

To understand wave propagation in waveguiding structures, it helps to have a simple heuristic picture of the phenomenon. Such a picture can be given by geometrical optics (Born and Wolf 1975, Chapter 3; Mickelson 1992, Chapter 5). As the reader will recall, the basic idea in geometrical optics is to expand Maxwell's equations for propagation in a slowly varying medium (on a scale of wavelengths). When such an expansion is valid, one finds that the optical wavefront can be broken up into local plane waves. In this way, one can solve reflection problems, for example, by applying Fresnel relations pointwise, as long as the "points" are larger than a few wavelengths of the optical wave. The Fresnel problem is illustrated in Figure 2.2. In the figure, the plane of incidence is the $x$-$z$ plane. As the interface is plane, were the wave incident in any plane other than the $x$-$z$ plane, one could perform a coordi-

8

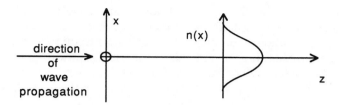

**FIGURE 2.1.**    Typical geometry used in this chapter, with $x$ the axis along which the index of refraction can vary, $y$ a perpendicular axis, and $z$ the direction of propagation of the incident electromagnetic disturbance.

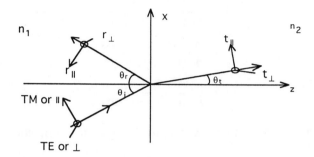

**FIGURE 2.2.**    Relative propagation and electric field polarization directions involved in the Fresnel problem for an interface between media with refractive indices $n_1$ and $n_2$.

nate transformation to put it in the $x$-$z$ plane. Also, from the symmetry of the plane interface, it is clear that the problem will have two electric field polarization eigenstates, one that is perpendicular to the plane of incidence and one which lies in the plane of incidence. Clearly, a wave incident in either of these states will remain in that state upon either reflection or transmission. This is not in general true for other polarization states. The state with its E-vector perpendicular to the plane of incidence is referred to as the *perpendicular* or *transverse electric* (TE) state. Clearly, in this state the electric vector will always remain transverse to the propagation direction. The other eigenstate will be the one in which the magnetic vector **H** remains perpendicular to the plane of incidence, requiring that the electric field lie in the plane of incidence. This electric field polarization state is referred to as the *parallel* or *transverse magnetic* state.

At this point, one could solve the Fresnel problem as a typical boundary value problem. This is a technique which will be applied repeatedly in this book. The point is that one knows the form of the solution in each of the two regions, 1 and 2. One can also derive directly from Maxwell's equations (as will be done in Section 2.2) the conditions that the electric and magnetic fields must satisfy at the interfaces between the media. If one can then satisfy these so-called boundary conditions, then one has a solution to the problem. The Fresnel problem is solved in such a manner in various places (see for example Mickelson 1992 or Born and Wolf 1975). The reflection and transmission coefficients for the problem can be written in the form (Mickelson, 1992)

$$t_\perp = \frac{2n_1 \cos \theta_i}{n_1 \cos \theta_i + n_2 \cos \theta_t} \tag{a}$$

$$t_\parallel = \frac{2n_1 \cos \theta_i}{n_2 \cos \theta_i + n_1 \cos \theta_t} \tag{b}$$

$$r_\perp = \frac{n_1 \cos \theta_i - n_2 \cos \theta_t}{n_1 \cos \theta_i + n_2 \cos \theta_t} \tag{c}$$

$$r_\parallel = \frac{n_2 \cos \theta_i - n_1 \cos \theta_t}{n_2 \cos \theta_i + n_1 \cos \theta_t} \tag{d} \quad (2\text{-}1)$$

where Equation (2-1) can be augmented by the law of reflection

$$\theta_i = \theta_r \tag{2-2}$$

and Snell's law

$$n_1 \sin \theta_i = n_2 \sin \theta_t \tag{2-3}$$

where $n_1$ and $n_2$ are the refractive indices of the two media, respectively, $\theta_i$ is the angle or incidence, $\theta_r$ is the angle of reflection, and $\theta_t$ is the transmission angle.

Here we are really only interested in one aspect of the relations of Equations (2-1)–(2-3)—that is, the phenomenon of total internal reflection (TIR). Clearly, if $n_1 > n_2$, there exists an incident angle $\theta_{TIR}$, defined by

$$\sin \theta_{TIR} = \frac{n_2}{n_1} \tag{2-4}$$

such that, for incident angles $\theta_i > \theta_{TIR}$, the sine of the transmitted angle becomes greater than one, which indicates that the cosine of the transmitted angle is expressible as

$$\cos \theta_t = i\sqrt{\sin^2 \theta_t - 1} = i\sqrt{\frac{n_1^2}{n_2^2} \sin^2 \theta_i - 1} \tag{2-5}$$

where $i$ is the square root of minus one. If one uses (2-5) in (2-1), one finds immediately that the reflection coefficients are of the form

$$r_\perp = e^{i\delta_\perp} \qquad \text{(a)}$$

$$r_\parallel = e^{i\delta_\parallel} \qquad \text{(b)} \quad (2\text{-}6)$$

and we see that the transmitted fields have the $z$ dependence $\exp\left[-kz\sqrt{n_1^2/n_2^2 \sin^2\theta_i - 1}\right]$, where $k$ is the propagation constant, which is equal to the angular frequency $\omega$ divided by the speed of light $c$, indicating that no energy is transmitted in the $z$ direction, despite there being finite values for the transmission coefficients. (This lack of energy flow follows from a calculation of the Poynting vector $\mathbf{S}$ using time harmonic dependence together with the $z$ dependence mentioned above and the relation (2-11a) between $\mathbf{E}$ and $\mathbf{H}$.)

We are now ready to apply geometrical optics to a waveguiding problem. Considering a slab guide as is depicted in Figure 2.3. Rays incident on the $n_1$-$n_2$ boundary at normal angles of less than $\theta_{\text{TIR}}$ will be partially transmitted and partially reflected by the interface. Such rays will be rapidly attenuated in propagating down the guide. Rays which have a normal angle greater than $\theta_{\text{TIR}}$, however, will be totally reflected back into the $n_1$ medium by the interface. The gap between the point at which the ray first impinges on the interface and the place where it leaves the interface is known as the *Goos-Hänchen* shift, which is due to the fact that the reflection coefficients in Equation (2-6) are complex numbers with unity magnitude. The unity magnitude gives the total internal reflection while the phase must be added to the propagating phase and therefore causes the resultant electromagnetic disturbance to appear to jump forward. It should be noted here that propagation in a continuously varying medium can also be analyzed geometrically via the one-dimensional paraxial ray equation, although the present author knows of no way one can get the Goos-Hänchen shift in that case. That there is such a shift even in the continuous case is due to the fact that there does exist a so-called turning point at which the index $n$ times the $\mathbf{k}$ vector's magnitude $k$ becomes less than the propagation constant and the solution becomes

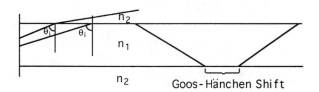

**FIGURE 2.3.**    Slab waveguide of index $n_1$ immersed in a medium of index $n_2$, with two ray paths for two different initial ray positions and angles.

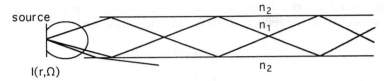

**FIGURE 2.4.**    Source with a specific intensity given by $I(r,\Omega)$ radiating into a slab wave-guide of a given numerical aperture.

damped. This corresponds to imaginary rays which can add phase as in the homogeneous slab case above.

The above described geometrical optics approach to guidance is actually more than a tool for understanding the guidance mechanism. It can also be used as a powerful computational aid for calculating such things as coupling efficiencies from incoherent sources into multimoded waveguides. Consider the situation depicted in Figure 2.4. Here, a partially coherent source whose irradiance is described by a specific intensity $I(\mathbf{r},\Omega)$ is radiating into the front facet of a slab waveguide. One can clearly represent the source in terms of a bundle of rays, the density of which is determined by the $\mathbf{r}$ dependence in $I(\mathbf{r},\Omega)$ and whose directions are determined by the $\Omega$ dependence of $I(\mathbf{r},\Omega)$. As the waveguide has a sharp cutoff in terms of ray angles, it can easily be determined which rays couple and which do not. Indeed, this analysis technique is the basic one of radiometry and will be taken up later in this book, in Chapter 5, Section 5 on optical fibers.

## 2.2 ELECTROMAGNETIC THEORY OF SLAB WAVEGUIDES

Here we wish to begin from Maxwell's equations to come to a solution of the problem depicted in Figure 2.1. We will write Maxwell's equations in the form

$$\nabla \times \mathbf{E}(\mathbf{r},t) = -\frac{\partial \mathbf{B}(\mathbf{r},t)}{\partial t} \tag{a}$$

$$\nabla \times \mathbf{H}(\mathbf{r},t) = \mathbf{J}(\mathbf{r},t) + \frac{\partial \mathbf{D}(\mathbf{r},t)}{\partial t} \tag{b}$$

$$\nabla \cdot \mathbf{D}(\mathbf{r},t) = \rho(\mathbf{r},t) \tag{c}$$

$$\nabla \cdot \mathbf{B}(\mathbf{r},t) = 0 \tag{d} \quad (2\text{-}7)$$

As is customary, let us start with the questionable assumption that the propagation is that of a monochromatic wave and therefore that one can take

$$\mathbf{E}(\mathbf{r},t) = \text{Re}[\mathbf{E}(\mathbf{r})e^{-i\omega t}] \qquad \text{(a)}$$

$$\mathbf{H}(\mathbf{r},t) = \text{Re}[\mathbf{H}(\mathbf{r})e^{-i\omega t}] \qquad \text{(b)}$$

$$\mathbf{J}(\mathbf{r},t) = \text{Re}[\mathbf{J}(\mathbf{r})e^{-i\omega t}] \qquad \text{(c)}$$

$$\rho(\mathbf{r},t) = \text{Re}[\rho(\mathbf{r})e^{-i\omega t}] \qquad \text{(d)} \quad \text{(2-8)}$$

but with the assurance that we can generate a more general disturbance by making the associations that

$$\mathbf{E}_\omega(\mathbf{r}) = \mathbf{E}\,(\mathbf{r}) \qquad \text{(a)}$$

$$\mathbf{E}_{\text{tot}}(\mathbf{r},t) = \int d\omega e^{-i\omega t} \mathbf{E}_\omega(\mathbf{r}) e^{i\phi(\omega)} \qquad \text{(b)} \quad \text{(2-9)}$$

where $\mathbf{E}_\omega(\mathbf{r})$ is the solution to Maxwell's equations for an angular frequency $\omega$, $\mathbf{E}_{\text{tot}}(\mathbf{r},t)$ is a polychromatic field, and the $\phi(\omega)$ must be determined from an initial condition such as a specification of the total field along a plane $z = 0$. Using (2-8) in (2-7), one finds that

$$\nabla \times \mathbf{E}(\mathbf{r}) = i\omega\mathbf{B}(\mathbf{r}) \qquad \text{(a)}$$

$$\nabla \times \mathbf{H}(\mathbf{r}) = \mathbf{J} - i\omega\mathbf{D}(\mathbf{r}) \qquad \text{(b)}$$

$$\nabla \cdot \mathbf{D}(\mathbf{r}) = \rho(\mathbf{r}) \qquad \text{(c)}$$

$$\nabla \cdot \mathbf{B}(\mathbf{r}) = 0 \qquad \text{(d)} \quad \text{(2-10)}$$

To simplify (2-10) further, we need to assume constitutive relations. At optical frequencies, it is always safe to assume that $\mathbf{B} = \mu_0\mathbf{H}$, as particles cannot move fast enough to set up circulating macroscopic currents at optical frequencies. Although there are no materials that are actually isotropic, let us follow the modal setup in Figure 2.1 and take that $\mathbf{D} = \epsilon\,\mathbf{E}$ with $\epsilon = \epsilon_0\, n^2(x)$. Further, for the moment at least, take $\rho$ and $\mathbf{J}$ both to be negligible. The assumption on $\mathbf{J}$ will be discussed at some length later in this chapter before the problems of loss and gain media are considered. With the simplification of this paragraph, the system of (2-10) can be written as

$$\nabla \times \mathbf{E}(\mathbf{r}) = i\omega\mu_0\mathbf{H}(\mathbf{r}) \qquad \text{(a)}$$

$$\nabla \times \mathbf{H}(\mathbf{r}) = -i\omega\epsilon\mathbf{E}(\mathbf{r}) \qquad \text{(b)}$$

$$\nabla \cdot (\epsilon\mathbf{E}(\mathbf{r})) = 0 \qquad \text{(c)}$$

$$\nabla \cdot \mathbf{H}(\mathbf{r}) = 0 \qquad \text{(d)} \quad \text{(2-11)}$$

In anticipation of assuming a definite propagation direction ($z$) that will break off vector symmetries, we write out the two curl equations (2-11a) and (2-11b) explicitly, to obtain

$$\frac{\partial E_z}{\partial y} - \frac{\partial E_y}{\partial z} = i\omega\mu_0 H_x \qquad \text{(a)}$$

$$\frac{\partial E_x}{\partial z} - \frac{\partial E_z}{\partial x} = i\omega\mu_0 H_y \qquad \text{(b)}$$

$$\frac{\partial E_y}{\partial x} - \frac{\partial E_x}{\partial y} = i\omega\mu_0 H_z \qquad \text{(c)}$$

$$\frac{\partial H_z}{\partial y} - \frac{\partial H_y}{\partial z} = i\omega\epsilon E_x \qquad \text{(d)}$$

$$\frac{\partial H_x}{\partial z} - \frac{\partial H_z}{\partial x} = i\omega\epsilon E_y \qquad \text{(e)}$$

$$\frac{\partial H_y}{\partial x} - \frac{\partial H_x}{\partial y} = i\omega\epsilon E_z \qquad \text{(f)} \quad \text{(2-12)}$$

One point can be made immediately: in the absence of other symmetries, (2-12) is a hopelessly coupled system. For example, $H_x$ will drive $E_z$ and $E_y$, while $E_z$ drives both $H_x$ and $H_y$, and $E_y$ drives both $H_x$ and $H_z$, where $H_z$ in turn drives $E_x$. All the field components are therefore, in general, coupled. Clearly more simplification is necessary before one can make any general statements about the system of (2-12). Since we are looking for modal solutions of (2-12) with $z$ directed propagation, an additional simplification is indeed available. Consider the situations depicted in Figure 2.5. For an electromagnetic disturbance to be a mode, it must retain its shape when being propagated forward. That is to say that the transverse dependence of the field cannot change, and the only effect of a translation along the propagation direction is to multiply the function by a phase factor. This is illustrated in Figure 2.5(a). As the medium under consideration must not change along the propagation direction if there are to be modes, then it makes no difference where we start our translation of length $L$. This is indicated by the depiction in Figure 2.5(b). It is therefore clear that the phase function must be linear in the translation distance, and therefore one could write

$$\mathbf{E}(\mathbf{r}) = \mathbf{E}(x,y,z) = \mathbf{E}(x,y)e^{i\beta z} \qquad \text{(a)}$$

$$\mathbf{H}(\mathbf{r}) = \mathbf{H}(x,y,z) = \mathbf{H}(x,y)e^{i\beta z} \qquad \text{(b)} \quad \text{(2-13)}$$

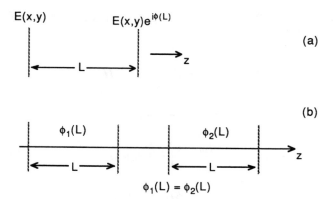

**FIGURE 2.5.**    Phase relations necessary for a disturbance to be an electromagnetic mode (a) the propagation effect (b) relation between two propagations of identical length.

giving

$$E(r,t) = Re[E(x,y)e^{i\beta z}e^{-i\omega t}] \qquad \text{(a)}$$

$$H(r,t) = Re[H(x,y)e^{i\beta z}e^{-i\omega t}] \qquad \text{(b)} \quad (2\text{-}14)$$

If the functions $E(x,y)$ and $H(x,y)$ can be chosen to be real (as we shall soon see they can in a lossless, slablike medium), then equation (2-14) indicates that the mode shapes $E(x,y)$ and $H(x,y)$ propagate forward like plane waves with flat fronts, and therefore, from the tracing of the ray paths, one can say that all the energy is propagated directly down the $z$ axis, the direction which is everywhere perpendicular to the equiphase surfaces, as illustrated in Figure 2.6, where several successive equiphase surfaces are plotted.

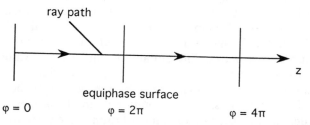

**FIGURE 2.6.**    Plane phase fronts predicted by Equation (2-14) for real mode shapes $E(x,y)$ and $H(x,y)$ where $\varphi$ is given by $\beta z - \omega t$.

Using (2-13) in (2-12), one immediately finds the system of equations (Marcuse 1982)

$$\frac{\partial E_z}{\partial y} - i\beta E_y = i\omega\mu_0 H_x \qquad \text{(a)}$$

$$i\beta E_x - \frac{\partial E_z}{\partial x} = i\omega\mu_0 H_y \qquad \text{(b)}$$

$$\frac{\partial E_y}{\partial_x} - \frac{\partial E_x}{\partial y} = i\omega\mu_0 H_z \qquad \text{(c)}$$

$$\frac{\partial H_z}{\partial y} - i\beta H_y = -i\omega\epsilon E_x \qquad \text{(d)}$$

$$i\beta H_x - \frac{\partial H_z}{\partial x} = -i\omega\epsilon E_y \qquad \text{(e)}$$

$$\frac{\partial H_y}{\partial x} - \frac{\partial H_x}{\partial y} = -i\omega\epsilon E_z \qquad \text{(f)} \quad \text{(2-15)}$$

which still appears to be a hopelessly coupled system. However, even at this level, the problem can be reduced considerably. By first eliminating $E_y$ and then $H_x$ from (2-15a) and (2-15e), one can express $H_x$ and $E_y$ as functions of only the longitudinal fields $E_z$ and $H_z$. A comparable treatment of $E_x$ and $H_y$ in (2-15b) and (2-15d) expresses these other two transverse fields $E_x$ and $H_y$ in terms of the longitudinal fields. By using the four relations for the transverse fields in terms of the longitudinal fields, one can derive (coupled) equations for $E_z$ and $H_z$ (Kurtz and Streifer 1969). In this sense, $E_z$ and $H_z$ serve as potentials for solving the complete problem. The equations for $E_z$ and $H_z$ are coupled second order equations, so that for each value for $\beta$ and $\omega$ there are four solutions—that is, four forward propagating polarization states. By reversing the sign of $\beta$, one could find another four backward travelling polarization states. As each of these solutions contain both $E_z$ and $H_z$, they are known as hybrid modes. The problem with solving the problem in the above outlined manner is that the coupled equation for $E_z$ and $H_z$ is in general a fourth order partial differential nonstandard eigenvalue equation with inhomogeneous coefficients. Later in this text, we shall see that such an equation can only be solved analytically for the case of the homogeneous dielectric rod. In other cases, the only solution method is numerical and does not shed much light on the nature of the solution. Here we wish to build up our solutions from simpler solutions.

Going back to our archetypical geometry of Figure 2.1, we recall that we were going to assume that there was no $y$ variation in the problem. Clearly, the modes of this structure must be of infinite extent in the $y$ dimension (and therefore excited by infinite sources). For such a situation, we can assume that the $y$ derivative must always be zero. Using this ansatz (assumption) in Equation (2-15), one immediately finds that the system of (2-15) decomposes into two systems of three equations each, which can be written in the form (Mickelson 1992, Chapter 2)

$$\frac{\partial H_z}{\partial x} = i\beta H_x + i\omega\epsilon E_y \qquad \text{(a)}$$

$$i\omega\mu_0 H_x = -i\beta E_y \qquad \text{(b)}$$

$$\frac{\partial E_y}{\partial x} = i\omega\mu_0 H_z \qquad \text{(c)} \quad (2\text{-}16)$$

and

$$\frac{\partial E_z}{\partial x} = i\beta E_x - i\omega\mu_0 H_y \qquad \text{(a)}$$

$$i\omega\epsilon E_x = -i\beta H_y \qquad \text{(b)}$$

$$\frac{\partial H_y}{\partial x} = -i\omega\epsilon E_z \qquad \text{(c)} \quad (2\text{-}17)$$

The structures of these two field systems are sketched in Figure 2.7. The results are actually not at all surprising following the discussion of the Fresnel coefficients of the opening paragraphs of this chapter. There it was noticed that the plane structure discussed gave rise to two natural polarization states, one with an always transverse $E$-field and one with an always transverse $H$-field. But the present problem being considered is essentially the Fresnel

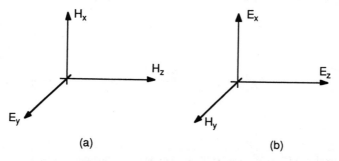

(a)                              (b)

**FIGURE 2.7.**    Field configurations which correspond to Equations (2-16) and (2-17).

problem turned on its side, and therefore it is not surprising that the polarization eigenstates are as they are illustrated in Figure 2.8. Indeed, the eigenstate defined by Equation (2-16) is generally referred to as the *TE polarization* and the state of equation (2-17) as the *TM polarization*.

It is reasonably easy to generate a single equation from the system of (2-16). Note that (2-16b) can be rewritten in the form

$$E_y = -\eta_{TE}H_x \qquad \text{(a)}$$

$$\eta_{TE} = \frac{\omega}{\beta_{TE}}\mu_0 = \frac{1}{n_{TE}}\sqrt{\frac{\mu_0}{\epsilon_0}} = \frac{\eta_0}{n_{TE}} \qquad \text{(b)} \quad \text{(2-18)}$$

where $n_{TE}$ is the effective index of refraction for the TE model, defined by $n_{TE} = \beta/k_0$, and $\eta_0$ is the impedance of free space. Equation (2-18a) is identical to the relation between **E** and **H** for a plane wave, with the index of the homogeneous region supporting the plane wave replaced by the effective index of the given mode. Therefore, TE and TM solutions have many of the attributes of plane wave solutions. Clearly, the fact that there are only two forward going and two backward going solutions is one of these attributes. In the case of both $x$ and $y$ index variation, we saw that there were twice as many solutions. Indeed, one could imagine modes which had a predominantly $x$-polarized $E$-field, yet a larger $\eta H_z$ than $E_z$. The symmetry of the slab geometry precludes these solutions but causes them to become degenerate with the TE and TM solutions. In this sense, TE and TM solutions are really made up of two hybrid modes whose degeneracy is broken when slab symmetry is broken.

Equation (2-16c) serves as a relation between the fields tangential to the index variation direction. This relation is of fundamental importance in cases where one considers an abrupt interface. In such cases, one knows that the tangential fields will be continuous across the interface (Born and Wolf 1975, Section 1.1.3). This is to say that $E_y$ and $H_z$ are continuous across the interface by using (2-16c), and one can translate that boundary condition

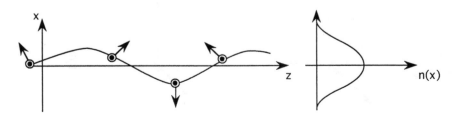

**FIGURE 2.8.** Possible ray path in a continuously varying medium whose index is plotted in the right-hand part of the diagram, indicating the polarization eigenstates.

into one in $E_y$ and $\partial E_y/\partial x$. That is, by using (2-16b) and (2-16c) in (2-16a), one can obtain the equation

$$\frac{\partial^2 E_y}{\partial x^2} + (k^2 - \beta^2)E_y = 0 \qquad (2\text{-}19)$$

subject to the auxiliary conditions for (2-16c) that

$$E_y, \frac{\partial E_y}{\partial x} \text{ are continuous} \qquad (2\text{-}20)$$

Similar reasoning leads to the TM system that

$$\frac{\partial H_y}{\partial x^2} + (k^2 - \beta^2)H_y = 0 \qquad (2\text{-}21)$$

with

$$H_y, \frac{1}{\epsilon} \frac{\partial H_y}{\partial x} \text{ continuous} \qquad (2\text{-}22)$$

Although the TE and TM boundary conditions are different, in many practical cases they reduce to the same thing. Consider a homogeneous slab as in Figure 2.3. If one considers the weakly guiding situation where

$$2\frac{n_1 - n_2}{n_1 + n_2} = \frac{2\Delta n}{n_1 + n_2} = \frac{\Delta n}{\bar{n}} \ll 1 \qquad (2\text{-}23)$$

where the first equality defines $\Delta n$ and the second defines $\bar{n}$, then it is clear that $\partial H_y/\partial x$ is approximately continuous, and therefore in this case there is a rough symmetry between the polarization states. It is also interesting that, in this weakly guiding case, regardless of the waveguide shape in terms of $x$ and $y$ variations, the modes will essentially always be (linearly polarized) TE and TM mode solutions. The reason for this is that, except in the case of perfect circular symmetry (see Figure 2.9), the modes of a structure will be linearly polarized, from symmetry. In a circular guide, the actual states should have some sort of circular symmetry, but given the weakly guiding limit, these states would just be degenerate linear combinations of linear states. A slight break from this perfect circular symmetry, however, would break the degeneracy of the linear states and make these the actual modes. In this sense, TE and TM modes are the modes of weakly guiding structures, at least in the

(a)                                    (b)

FIGURE 2.9.    Polarization of the modes in (a) a slightly elliptical waveguide and (b) a perfectly circular waveguide.

somewhat unrealistic case where we consider linear, isotropic media. We shall see much more on this point when we discuss fibers in Chapter 5.

We now want to consider solving the eigenvalue problem outlined in Equations (2-19) and (2-20). Again, we wish to consider a case such as in Figure 2.1. This is what we will do in the next section. First, we wish to point out some salient features of the problem by considering the asymmetric slab waveguide depicted in Figure 2.10.

The TE modes of the asymmetric slab must satisfy Equation (2-19) and, in each region and at the two boundary interfaces, must satisfy both conditions of (2-20). Now for a mode to be guided, it must have a finite percentage of its energy propagating within region II of the structure of Figure 2.10. Now as the solutions to the equations

$$\frac{\partial^2 \psi}{\partial x^2} + \lambda \psi = 0 \tag{2-24}$$

are of the form

$$\psi(x) = A \cos \sqrt{\lambda}\, x + B \sin \sqrt{\lambda}\, x, \qquad \lambda > 0 \qquad \text{(a)}$$

$$\psi(x) = A e^{\sqrt{|\lambda|}x} + B e^{-\sqrt{|\lambda|}x}, \qquad \lambda < 0 \qquad \text{(b)} \quad (2\text{-}25)$$

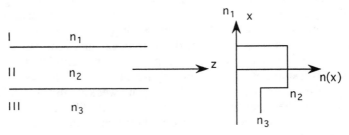

FIGURE 2.10.    Layers of an asymmetric slab waveguide, with its index plotted in the right of the figure.

one can determine the solution to (2-19) throughout space by just applying (2-20) and the condition for a mode to be guided, as was stated above. The argument is as follows: As the mode fields we wish to find must be in some manner normalized in order to carry finite power in region 2, they must be of finite transverse extent in $x$. In order to satisfy this requirement, we see that the solutions in regions I and III must be of the form

$$\psi_i(x) = A_i e^{\kappa_i x} + B_i e^{-\kappa_i x} \qquad (2\text{-}26)$$

where $i = 1,2$, $A_i$ and $B_i$ are constants to be determined from the finiteness and continuity conditions (2-20), and

$$\kappa_i = \sqrt{\beta^2 - k_i^2} \qquad \text{(a)}$$

$$k_i^2 = k_0 n_i^2 = \frac{2\pi}{\lambda} n_i^2 \qquad \text{(b)} \quad (2\text{-}27)$$

Clearly, the condition of finiteness in region II will also require that $A_1 = B_3 = 0$. The continuity conditions are also going to require that the solution in region II be of the form

$$\psi_{II}(x) = A_2 \cos \gamma_2 x + B_2 \sin \gamma_2 x \qquad (2\text{-}28)$$

with $\gamma_2$ defined by

$$\gamma_2 = \sqrt{k_2^2 - \beta^2} \qquad (2\text{-}29)$$

Figure 2.11 serves to illustrate why the solution in region II must take the form of (2-28) and not that of (2-26). For $n_2 \neq n_1$, $n_2 \neq n_3$, there would be no way to match both the functions and their derivatives at the boundaries. This is not true for the solution in (2-28), as we shall presently see.

The finiteness condition that $A_1 = B_3 = 0$ leaves one with four constants, $A_2$, $A_3$, $B_1$, and $B_2$. The conditions of (2-20) will yield four equations. These

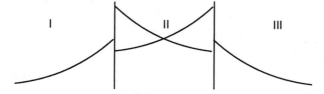

**FIGURE 2.11.**    The solutions in the various regions in the case where the solution in region II matched the forms of those in regions I and III.

equations are expressible in the form

$$A_1 e^{\kappa_1 x_{12}} + B_1 e^{-\kappa_1 x_{12}} = A_2 \cos \gamma_2 x_{12} + B_2 \sin \gamma_2 x_{12} \quad \text{(a)}$$

$$\kappa_1 A_1 e^{\kappa_1 x_{12}} - \kappa_1 B_1 e^{-\kappa_1 x_{12}} = -\gamma_2 A_2 \sin \gamma_2 x_{12} + \gamma_2 B_2 \cos \gamma_2 x_{12} \quad \text{(b)}$$

$$A_2 \cos \gamma_2 x_{23} + B_2 \sin \gamma_2 x_{23} = A_3 e^{\kappa_3 x_{23}} + B_3 e^{-\kappa_3 x_{23}} \quad \text{(c)}$$

$$-\gamma_2 A_2 \sin \gamma_2 x_{23} + \gamma_2 B_2 \cos \gamma_2 x_{23} = \kappa_3 A_3 e^{\kappa_3 x_{23}} - \kappa_3 B_3 e^{-\kappa_3 x_{23}} \quad \text{(d)} \quad \text{(2-30)}$$

where $x_{12}$ is the coordinate of the interface between regions I and II and $x_{23}$ is the coordinate of the interface between regions II and III. Clearly $A_1$ and $B_3$ could have been set to zero in (2-30) but have been left in for the moment, for a reason we shall soon see. Clearly, Equations (2-30) form a system of four linear homogeneous equations in four unknowns. Such equations will only have a solution if the determinant of coefficients of the variables is zero. This determinant will not in general be zero but will place a requirement on the propagation constant $\beta$ of Equation (2-13). Actually, (2-26) and (2-28) already place the requirement that

$$k_0 n_2 > \beta > \max \lceil k_0 n_1, k_0 n_3 \rceil \tag{2-31}$$

where max denotes the maximum function. Were $\beta$ to exceed $k_0 n_2$, the situation would be like that of Figure 2.11 and the boundary conditions could not be satisfied. One could also think of this in terms of $\beta$ as being the z-component of the $k$ vector. The largest total $k$ vector possible in a guiding layer with index $n_2$ is $k_0 n_2$, and this only if the wave is completely confined within the medium, that is, as if it is a plane wave in a homogeneous medium of index $n_2$. Clearly, the z-component of the $k$ vector cannot exceed the total $k$ vector. In fact, the largest value the $\beta$ could take on would be $k_0 n_2$, and that would be for a z-directed plane wave in a homogeneous medium of index $n_2$. For the guiding case, $\beta$ must clearly be less than this. The other side of the inequality has a comparable interpretation. Clearly, if $\beta$ becomes less than, for example, $k_0 n_3$, then the solution in region III becomes of the form of Equation (2-28). As this solution is propagating, it means that the wave is no longer bound in region II but propagates in the semi-infinite region III. In terms of vector components, when $\beta = k_0 n_3$, then $\beta$ corresponds to the z-component of a plane wave propagating in region III. As region III is semi-infinite, it can support plane wavelike solutions. As $\beta$ becomes less than $k_0 n_3$, this plane wave propagates at ever increasing angles from the z-axis into region III. Because these solutions with $\beta = k_0 n_3$ indicate plane wave solutions in region III, modes with such propagation constants are generally referred to as *cladding modes,* because they exist in the cladding and not in the guide. We now see that setting the determinant of the coefficient matrix of (2-30) equal to zero has the effect of pinpointing $\beta$ values within the allotted range of the inequality of (2-31).

One can attempt to manipulate Equations (2-30) directly to obtain the dispersion relation (equation for determining $\beta$) and the one-parameter string which will define $A_2$, $A_3$, $B_1$, and $B_2$. (A one-parameter string is used because, after $\beta$ is determined, there are only three equations left to determine the four constants, and therefore they all have to be determined in terms of a normalization constant.) What we wish to do now, however, is to develop a general technique for approaching such problems. This technique can yield the solution in a prescribed manner, it gives us a notation that relieves us from writing out a lot of algebra, and it will be used repeatedly as the course continues. The basic idea behind the technique is that the two sets of boundary conditions of Equations (2-30) can be written as (not quite unique) matrix equations. (This is done for propagating waves in both Mickelson 1992, Chapter 2 and Born and Wolff 1975, Chapter 1. For the bound mode formulation, see Charczenko and Mickelson, 1989.) Defining the following matrices

$$
R_1 = \begin{pmatrix} e^{\kappa_1 x_{12}} & e^{-\kappa_1 x_{12}} \\ \dfrac{\kappa_1}{k_0} e^{\kappa_1 x_{12}} & -\dfrac{\kappa_1}{k_0} e^{-\kappa_1 x_{12}} \end{pmatrix},
$$

(a)

$$
L_2 = \begin{pmatrix} \cos \gamma_2 x_{12} & \sin \gamma_2 x_{12} \\ -\dfrac{\gamma_2}{k_0} \sin \gamma_2 x_{12} & \dfrac{\gamma_2}{k_0} \cos \gamma_2 x_{12} \end{pmatrix}
$$

(b)

$$
R_2 = \begin{pmatrix} \cos \gamma_2 x_{23} & \sin \gamma_2 x_{23} \\ -\dfrac{\gamma_2}{k_0} \sin \gamma_2 x_{23} & \dfrac{\gamma_2}{k_0} \cos \gamma_2 x_{23} \end{pmatrix},
$$

(c)

$$
L_3 = \begin{pmatrix} e^{\kappa_3 x_{23}} & e^{-\kappa_3 x_{23}} \\ \dfrac{\kappa_3}{k_0} e^{\kappa_3 x_{23}} & -\dfrac{\kappa_3}{k_0} e^{-\kappa_3 x_{23}} \end{pmatrix}
$$

(d)            (2-32)

where $R_i$ ($L_i$) denotes the matrix for the right(left)-hand side of the $i^{\text{th}}$ medium, and the vectors

$$
V_i = \begin{pmatrix} A_i \\ B_i \end{pmatrix} \tag{2-33}
$$

One can express Equations (2-30) in the form

$$
R_1 V_1 = L_2 V_2 \tag{a}
$$

$$
R_2 V_2 = L_3 V_3 \tag{b (2-34)}
$$

where the situation is schematically depicted in Figure 2.12. It is clear that

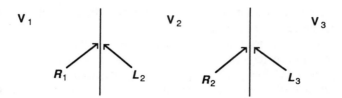

**FIGURE 2.12.** The use of matrices to relate the coefficient vectors in the three regions of Figure 2.10.

one can use (2-34) to relate $V_1$ and $V_3$. Clearly, one can write that

$$V_1 = R_1^{-1} L_2 V_2 \tag{2-35}$$

and

$$V_2 = R_2^{-1} L_3 V_3 \tag{2-36}$$

to obtain

$$V_1 = R_1^{-1} L_2 R_2^{-1} L_3 V_3 = R_1^{-1} M_2 L_3 V_3 = T V_3 \tag{2-37}$$

where Equation (2-37) should serve as a defining equation for the characteristic matrix $M_2$ and the transmission matrix $T$. Writing out (2-37) in explicit detail, one finds that

$$\begin{pmatrix} A_1 \\ B_1 \end{pmatrix} = \begin{pmatrix} T_{11} & T_{12} \\ T_{21} & T_{22} \end{pmatrix} \begin{pmatrix} A_3 \\ B_3 \end{pmatrix} \tag{2-38}$$

From the finiteness condition that $A_1 = B_3 = 0$, it becomes immediately obvious that the dispersion relation (between $\beta$ and $k_0$) becomes

$$T_{11} = 0 \tag{2-39}$$

and that, following solution of (2-39) for $\beta$, the required $\beta$ value could be backsubstituted into (2-32) and then $A_2$, $A_3$, and $B_2$ expressed in terms of $B_1$ through the use of (2-35) and (2-36). A normalization condition of the form

$$\int_{-\infty}^{\infty} |\psi(x)|^2 dx = N_B \tag{2-40}$$

can then be used to find the $B_1$ value. This normalization will be carried out in the following paragraphs.

It is now possible to study the solutions of (2-39) for the asymmetric slab waveguide. Before doing this, however, we wish to present the generalization of the matrix method to the archetypical problem of Figure 2.1. As we shall soon see, formally at least, the solution of a general case is as simple as that of the asymmetric slab. Clearly, one can make a staircase approximation to a continuous index distribution by breaking it up into constituent slabs. In this sense, one can easily reduce the problem illustrated in Figure 2.1 into an equivalent $N$ slab problem, as illustrated in Figure 2.13. Clearly, the solutions in regions I and $N$ must be dying exponentials, and therefore $A_1$ and $B_N$ are determined and must be zero. There is a bit of a problem in the intervening layers as, a priori, one does not know in which regions $\beta > k_0 n_i$ and in which $\beta < k_0 n_i$. This is not a fundamental problem, however, as one can assume a starting range on $\beta$, search that range, and then proceed to another range. Let us assume that we have specified a range, and therefore we know all of the $R$'s and $L$'s. One can therefore write that

$$R_i V_i = L_{i+1} V_{i+1}, \qquad i = 1, \ldots , \text{N-1} \tag{2-41}$$

Proceeding as in the asymmetric slab case, one can write that

$$V_i = R_i^{-1} L_{i+1} V_{i+1} \tag{2-42}$$

and

$$V_1 = R_1^{-1} L_2 R_2^{-1} \cdots L_{N-1} R_N^{-1} L_N V_N$$
$$= R_1^{-1} M_2 \cdots M_{N-1} L_N V_N = T V_N \tag{2-43}$$

where (2-43) also serves to define the characteristic matrices $M_i$ as well as the transmission matrix T. As before, it is easy to see that $\beta$ can be determined from the condition $A_1 = B_N = 0$, which gives

$$T_{11} = 0 \tag{2-44}$$

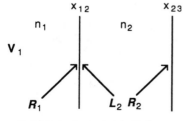
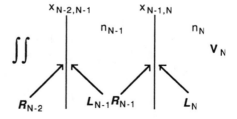

**FIGURE 2.13.**    $N$ slab medium.

and that all the coefficients $A_2, \ldots, A_N, B_2, \ldots, B_{N-1}$, can be determined in terms of $B_1$ from backsubstituting the $\beta$ solutions of (2-44) into (2-42). Formally at least, the general case is the same as the asymmetric slab case.

We now wish to consider a symmetric slab waveguide in some detail. It is not hard to show that the condition $T_{11} = 0$ for the asymmetric slab waveguide of Figure 2.10 reduces to

$$\tan \gamma L = \frac{\gamma L(\kappa_1 + \kappa_3)L}{\gamma^2 L^2 - \kappa_1 \kappa_3 L^2} = F(\gamma L) \tag{2-45}$$

where the equality sign serves to define the function $F$ and $L$ is the thickness of the guiding layer. Limiting present consideration to the symmetric slab case, one can take $\kappa = \kappa_1 = \kappa_3$ in (2-45) to obtain

$$\tan \gamma L = \frac{2\gamma L \, \kappa L}{\gamma^2 L^2 - \kappa^2 L^2} = F(\gamma L) \tag{2-46}$$

Perhaps the best way to understand the contents of (2-46) is to display it graphically, as is done in Figure 2.14. Here, the branches of the tangent function are plotted versus $\gamma L$ on the same set of axes as the function of $F(\gamma L)$. The intersection points between these two functions are the possible solutions of the dispersion relation. Recalling the definitions of $\kappa$ and $\gamma$ of Equations (2-27) and (2-29), one can show that

$$\kappa^2 = k_0^2(n_2^2 - n_1^2) - \gamma^2 \tag{2-47}$$

and therefore that $F(\gamma L)$ only remains real for

$$\gamma L < V \tag{2-48}$$

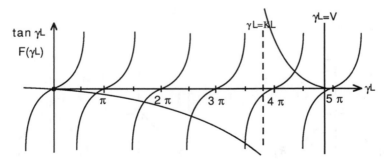

FIGURE 2.14.    Graphical representation of the functions of the left and right-hand sides of Equation (2-46).

where the number $V$ is defined by

$$V = k_0 L \sqrt{n_2^2 - n_1^2} \tag{2-49}$$

Such a line is sketched in Figure 2.14. Clearly, as $V$ decreases, there will be fewer intersection points. It is important to note, however, the $\kappa L$ is always less than $V$ (from (2-27) and (2-49)) and further $F(\gamma L = V) = 0$ from (2-47), and therefore, for the symmetric structure considered here, there will always be at least two intersection points, one at the origin and one where the upper branch of $F$ intersects an upper branch of the tangent function.

We now wish to discuss the properties of the first few solutions of (2-46). The one at the origin is a bit of a strange one. This solution has $\gamma L = 0$ and therefore $\beta = k_0 n_2$. This is the solution in which there is a uniform $z$-directed wave in region II. As Figure 2.15 illustrates, there is no way that this solution can satisfy the boundary conditions of (2-30). Such a solution could only exist at zero frequency where there would be no decay in regions I or III. A uniform D.C. field throughout space is of no interest to us here. We now wish to look at the next solution as it is depicted in Figure 2.14. This solution has $\gamma L$ approaching $\pi$. This means that almost one period of the cosine function fits into region II. The situation is as depicted in Figure 2.16, where the first three modal structures and relative $\beta$ values are plotted. As mentioned above, the first $\gamma L$ value is roughly $\pi$, and therefore this fundamental mode can have no zero crossing as it does not generate enough phase across its propagating region. The second solution has a $\gamma L$ slightly less than $2\pi$ and therefore will have a null in the center of the guiding layer. This extra phase is generated by lowering the $\beta$ value relative to the value of the fundamental $\beta$. Using the interpretation that $\beta$ is the $z$-component of the $k$ vector, one notes that this second order mode will have a larger transverse component of $k$ and therefore in the picture of Figure 2.3 would bounce back and forth between the guide walls at a higher angle than that of the lowest order mode. As with the third order mode of the plot of Figure 2.16, the same comments that

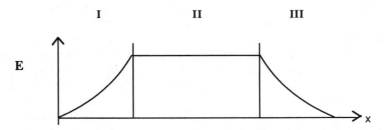

**FIGURE 2.15.**    Sketch indicating the spurious nature of the solution of the dispersion relation with $\gamma L = 0$.

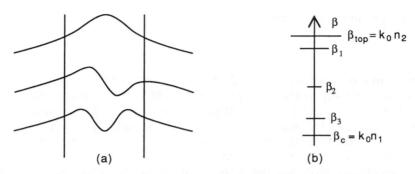

**FIGURE 2.16.** The first three modal solutions (a) and the relative $\beta$ values of these solutions (b) for a situation as depicted by the graphical dispersion relation of Figure 2.14.

applied to the second mode will apply to all the higher ones. Each higher mode will have an extra null and lowered $\beta$, giving a higher effective ray angle. This qualitative explanation will of course apply also to asymmetric slab guides. It is also clear that, for a $V$ value $N\pi < V < (N + 1)\pi$, there will be $(N + 1)$ propagating, bound modes in the guide. Therefore, from simple index and thickness measurements, one can determine a priori how many modes will propagate in the guide.

An important point to be remembered in interpreting the solutions of the dispersion relation of Equation (2-46) and Figure 2.14 is that all the results are strongly dependent on the operating frequency $f = \omega/2\pi$. $V$, for example, is linearly dependent on $\omega$, and therefore the number of modes is frequency dependent. The propagation constant $\beta$, which determines the phase velocity of propagation $v_p$ through the relation $v_p = \omega/\beta$ is also frequency dependent, although in a reasonably complicated way. As real optical sources are of finite bandwidth, pulses launched from such a source will have a tendency to expand with propagation distance as each frequency component making up a wave packet will travel at a slightly different velocity, if only due to the difference in $\beta$ values. This effect is referred to as *waveguide dispersion*. This dispersive effect is compounded by the fact that the $n$'s are also functions of frequency. The frequency dependence of $n$ is often referred to as *material dispersion*. From the form (2-40), it is clear that material and waveguide dispersion are inextricably tangled, although they often are referred to separately. Perhaps the important point here is, however, that planar structured devices such as integrated optic devices are generally short in length (millimeters), whereas optical fibers are generally long (tens of kilometers). Therefore, it is generally true that one needs to take dispersion into account in fiber propagation, but not usually in integrated optical waveguide propagation. For this reason, detailed discussion of dispersion will be left for inclusion in Chapter 5 on fibers. For resonant interactions between wave and material, dispersion can be important over even micron distances. Such interaction is the material of Chapters 3 and 4.

The one remaining issue of the mode problem to be discussed here is that of mode normalization. The basic idea is that any electric field distribution propagating down the guide must be expressible in the form

$$\mathbf{E}(\mathbf{r},t) = \mathrm{Re}\left[\sum_n a_n \psi_n(x) e^{i\beta_n z} e^{-i\omega t}\right] \tag{2-50}$$

where the $\psi_n$'s are the vector modes of the structure. In general, these modes will include bound modes as well as the cladding or radiation modes. The radiation modes, however, should be heavily damped in the core region (due to the fact that they are propagating away from the core), and therefore, far enough away from a coupling point, their contribution to the total field should be negligible. Practically, therefore, one can take (2-50) as being a sum over guided modes if one considers long enough waveguides. Now the question of mode normalization is closely tied to the question of how to interpret what the $a_n$'s mean. Here we want to make a normalization such that $a_n^* a_n$ gives the power propagating in the $n^{\text{th}}$ mode. This means that modes $m$ and $n$ with the relation $a_m^* a_m = a_n^* a_n$ will be carrying the same power. Other normalizations are also possible but lead to different interpretations of the $a_n$'s.

Consider the complex Poynting vector $\mathbf{S}_n$ corresponding to the power carried in a mode $n$. One can write the complex Poynting vector in the form (Jackson 1975, Chapter 6)

$$\tilde{\mathbf{S}}_n = \frac{1}{2}(\mathbf{E}_n \times \mathbf{H}_n^*) \tag{2-51}$$

If one expresses the $y$ component of the electric field of the $n^{\text{th}}$ TE mode in the form

$$E_{yn}(\mathbf{r}) = a_n \psi_n(x,y) e^{i\beta_n z} \tag{2-52}$$

where here the $\psi_n$ is a scalar TE mode, one finds from (2-16) that the $\mathbf{H}_n$ vector can be written as

$$\mathbf{H}_n(\mathbf{r}) = -\frac{1}{\eta_n} E_{yn}(\mathbf{r})\hat{\mathbf{e}}_x - \frac{i}{\omega\mu_0}\frac{\partial E_{yn}}{\partial x}\hat{\mathbf{e}}_z \tag{2-53}$$

where $\eta_n$ is the impedance of the $n^{\text{th}}$ mode. For this case, the complex

Poynting vector will be given by

$$\hat{S}_n = +\frac{1}{2\eta_n} a_n^* a_n \psi_n^2(x)\hat{e}_z - \frac{i}{\omega\mu_0} a_n^* a_n \psi_n(x)\frac{\partial\psi_n(x)}{\partial x}\hat{e}_x \qquad (2\text{-}54)$$

Not surprisingly, the Poynting vector has two components. Therefore, the energy propagation will follow a zigzag path much like a smoothed version of that illustrated in Figure 2.3. That this is the case is clear from the fact that the $\hat{e}_x$ component of S is purely imaginary and therefore has a zero time average. (The time averaged Poynting vector is just the real part of the complex Poynting vector.) The only way that this is possible is if the $x$-component oscillates with a period that is half the optical period. The real $z$-component indicates a constant rate of energy transfer along the $z$-axis.

Now the power crossing a closed surface $A$, $P_A$, is given by

$$P_A = \int_A \text{Re}[\tilde{S}] \cdot dA \qquad (2\text{-}55)$$

where $dA$ is the outward pointing unit vector of $A$ and where the $\text{Re}(\tilde{S})$ operation is equivalent to the operation of taking the time average of the real Poynting vector under the assumption that the wave is "plane wave like." (Basically, the wave must be a propagating wave.) If one takes $A$ to be a plane at $z = $ constant (and its closure at infinity), one finds the expression

$$P_A = \frac{a_n^* a_n}{2\eta_n} \int |\psi_n(x)|^2 \, dx \qquad (2\text{-}56)$$

Now, a sensible way to normalize seems to be to have the power in each mode be directly proportional to the $a^*a$ for the mode, as was discussed above. Setting the proportionality constant equal to one would yield that

$$\frac{a_n^* a_n}{2\eta_n} \int |\psi_n(x)|^2 \, dx = a_n^* a_n \qquad (2\text{-}57)$$

and therefore that the normalization condition should be

$$\int |\psi_n(x)|^2 \, dx = 2\eta_n = \frac{2\eta_0}{n_n} \qquad (2\text{-}58)$$

where the effective mode index $n_n$ is given by

$$n_n = \frac{\beta_n}{k_0} \qquad (2\text{-}59)$$

With the normalization of (2-58), one can fix the $B_n(B_3)$ in the $\mathbf{V}_n(\mathbf{V}_3)$ of Equations (2-43) and (2-37) and thereby have a complete and unique solution for each of the guide modes.

## 2.3 EXTENSIONS TO GUIDES WITH LOSS OR GAIN

Although we shall soon see that there is a relatively easy way to stick gain and loss into the dielectric constant, we feel that it is necessary to discuss the meaning of this substitution in some detail before making it, as there is no single, straightforward explanation of the substitution's meaning. Further, as we shall soon be attempting to understand the dynamical behavior of gas and semiconductor media, now is a good time to begin thinking about some of the processes involved in guiding materials. The development will proceed by first discussing conduction effects in metals and then the process of dielectric absorption before giving a qualitative discussion of semiconductor gain and loss processes.

Generally, all loss processes are modeled as being due to a conduction current $\mathbf{J}$ which is proportional to an electric field $\mathbf{E}$. This corresponds to currents in a metal and is depicted schematically in Figure 2.17. The application of an electric field across a metallic structure will cause the loosely bound conduction electrons to begin to move. The motion, however, will be strongly damped by rapid (each $10^{-7}$ sec) collisions with other electrons, lattice points (or the fields emanating therefrom), phonons, etc. Therefore, the equation of motion for such a charge could be given by

$$m_e \frac{d^2x}{dt^2} + \gamma m_e \frac{dx}{dt} = -eE_x(t) \tag{2-60}$$

for a one-dimensional motion. Now, assuming a time harmonic field at $f = \omega/2\pi$, one can find the amplitude component of the motion of the field

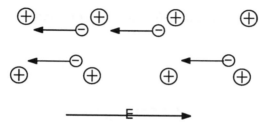

**FIGURE 2.17.**    Lattice with conduction electrons which move under the application of an electric field **E**.

at the frequency $\omega$, $x(\omega)$, by

$$x(\omega) = \frac{eE_{x_0}}{m_e\omega^2 + i\omega\gamma m_e} \tag{2-61}$$

where

$$x(t) = \text{Re}[x(\omega)e^{-i\omega t}] \tag{2-62}$$

Now, in general, an electron current can be expressed in the form

$$\mathbf{J}(t) = -Nev(t) \tag{2-63}$$

where $n$ is an electron density. If one were to write that $J_x(\omega)$ is the coefficient of $e^{-i\omega t}$ for time harmonic driving and that $v(\omega) = -i\omega x(\omega)$, one can express $J_x(\omega)$ in the form

$$J_x(\omega) = \frac{iNe^2 E_{x_0}}{m_e\omega + i\gamma m_e} = \sigma(\omega)E_{x_0} \tag{2-64}$$

where

$$\sigma(\omega) = \frac{Ne^2}{\gamma m_e - im_e\omega} \tag{2-65}$$

Equation (2-64) is essentially a one-dimensional statement of Ohm's law, which for isotropic media can clearly be generalized to

$$\mathbf{J}(\omega) = \sigma(\omega)\,\mathbf{E}(\omega) \tag{2-66}$$

Now, recalling that

$$\nabla \times \mathbf{H} = \mathbf{J} + \frac{\partial \mathbf{D}}{\partial t} \tag{2-67}$$

and substituting $\mathbf{D} = \epsilon\,\mathbf{E}$ and (2-66) in (2-67), one finds that

$$\nabla \times \mathbf{H} = -i\omega\,\hat{\epsilon}\,\mathbf{E} \tag{2-68}$$

where

$$\hat{\epsilon} = \epsilon + i\,\frac{\sigma(\omega)}{\omega} \tag{2-69}$$

is the complex refractive index.

To help understand the physical meaning of the complex index, let us consider two limits. One could easily consider a low frequency limit in which the period of the driving field were so long that the electron experienced many collisions before reversing its direction. It is safe to assume that in this limit $\gamma \gg \omega$, and therefore one could write that

$$\sigma \approx \frac{e}{\gamma} = \sigma(0) \tag{2-70}$$

Generally, the definition of the index of refraction is $\epsilon = n^2 \epsilon_0$, so here one could define a complex index of refraction $\hat{n}$ such that

$$\hat{\epsilon} = \hat{n}^2 \epsilon_0 \tag{2-71}$$

which gives one the relations that

$$n_r^2 - n_i^2 = \frac{\epsilon_r}{\epsilon_0} \tag{a}$$

$$n_r n_i = \frac{\sigma(0)}{2\omega\epsilon_0} \tag{b} \quad (2\text{-}72)$$

Assuming plane wave propagation in a material with complex index $\hat{n}$ yields the relation that

$$E(z) = E_0 e^{ik_0 \hat{n} z} = E_0 e^{ik_0 n_r z} \, e^{-k_0 n_i z} \tag{2-73}$$

Equation (2-73) tells us that the imaginary part of the index acts as an extinction coefficient. Equation (2-72) tells us that $\sigma(0)$ is the driving force behind this extinction as $\sigma(0) \to 0$ gives a real index, while $\sigma(0) \to \infty$ gives us a relation like $n_r \sim n_i$. If $n_r \sim n_i$, then the wave is damped to its $1/e$ point in a single period, which is indeed rapid damping. The energy tapped from the wave goes into the kinetic energy of the conduction electron gas in this model.

A second limit to consider is the one in which the driving period is so short that the electron is not scattered before the field reverses its motion. The situation is as depicted in Figure 2.18, where it is assumed that the field is so high frequency that the electron does not even traverse a single (roughly 5 Å) lattice site. In such a case, one can ignore $\gamma$ with respect to $\omega$ and obtain

$$\hat{\epsilon} = \epsilon + \frac{ne}{m_e \omega^2} = \epsilon_r \tag{2-74}$$

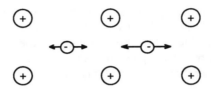

**FIGURE 2.18.**    The motion of conduction electrons in a very high frequency field.

which is a real index. Indeed, without scattering there is no place for energy to be lost, no channel of transfer, and therefore $n_i = 0$. Indeed, this is the case for metals, which generally begin becoming transparent in the optical or ultraviolet portions of the spectrum.

Before discussing absorption, it is of interest to see where the breakpoint between low and high frequency falls. Clearly, when the lattice spacing $a_0$ divided by the average drift velocity $v_d$ becomes of the order of magnitude of the electromagnetic period $T_{em}$, something must be happening. Using values of $a_0 \sim 5$ Å and $v_d \sim 10$ m/sec, one finds that

$$T_{em} = \frac{a_0}{v_d} = 5 \times 10^{-11} \text{ sec} \qquad (2\text{-}75)$$

which corresponds to a frequency of roughly 20 GHz. This means that radio waves are probably low frequency and optical waves are high frequency. What actually happens in this microwave/millimeter region is an open question, as it becomes a complicated probability problem in which quantum mechanical corrections can become very important.

Now we have seen that, in the case of conduction current, we can take losses into account by sticking a loss factor into the dielectric constant. What if the loss is absorptive? The standard model to employ for dielectric absorption is the harmonic oscillator model, which is depicted in Figure 2.19. The

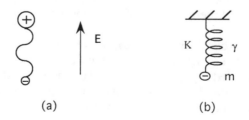

(a)                                (b)

**FIGURE 2.19.**    The harmonic oscillator model for the valence atoms in a dielectric where in (b) the lattice point is assumed to be stationary (large mass).

electron is considered to be attached by a spring of spring constant $K$ and friction force constant $\gamma$ to an assumed infinitely heavy nucleus. The effective mass of the electron can be varied in this model to give the correct oscillator strength. Considering a one-dimensional model for the moment, one can write the equation for the position $x(t)$ of the electron in the form

$$m\frac{d^2x(t)}{dt^2} + \gamma m\frac{dx}{dt} + Kx(t) = -eE(t) \tag{2-76}$$

As in the last section one assumes time harmonic variation of the field. One can then solve for the coefficient $x(\omega)$ of the time harmonic factor to yield

$$x(\omega) = \frac{-eE_0}{m(\omega_r^2 - \omega^2 - i\gamma\omega)} \tag{2-77}$$

where $E_0$ is the field strength and $\omega_r^2 = K/m$. The real and imaginary parts of $x(\omega)$ are plotted in Figure 2.20. In general, the motion of an electron will be due to the sum of many resonances and therefore can be written in the form

$$x(\omega) = \sum_k f_k \frac{E_0}{\omega_{rk}^2 - \omega^2 - i\gamma\omega} \tag{2-78}$$

where $f_k$ is the oscillator for the $k^{th}$ resonance at $\omega_{rk}$. We now wish to relate this atomic motion to the refractive index of the medium. The atomic dipole moment $\alpha(\omega)$ at angular frequency $\omega$ can be defined by the equation

$$\alpha(\omega)E_0 = -\frac{ex(\omega)}{\epsilon_0} = (\alpha_r + i\alpha_i)E_0 \tag{2-79}$$

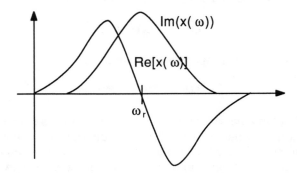

**FIGURE 2.20.**    Plots of the real and imaginary parts of $x(\omega)$.

and the index of refraction $n(\omega)$ can be defined in terms of $\alpha$ and the atomic dipole density $N$ by

$$n(\omega) = \sqrt{\frac{1 + \dfrac{2N\alpha}{3}}{1 - \dfrac{N\alpha}{3}}} = n_r + in_i \qquad (2\text{-}80)$$

From Figure 2.20 and Equations (2-78)–(2-80), it is clear that, far from a resonance, $n(\omega)$ will be real valued as will $\alpha$ and $x(\omega)$. Near the center of a resonance, however, the real part of $\alpha$ can be roughly zero, yielding for Equation (2-81)

$$n(\omega = \omega_{rk}) = \sqrt{\frac{1 + i\dfrac{2}{3}N\alpha_i}{1 - i\dfrac{N\alpha_i}{3}}} \qquad (2\text{-}81)$$

If the medium is rarified or the transition is weak, then $N\alpha_i \ll 1$ and one can express

$$n(\omega = \omega_{rk}) \sim 1 + i\frac{N\alpha_i}{2} \qquad (2\text{-}82)$$

yielding a plane wave propagation term of

$$E(z) = E(0)e^{ik_0 z}e^{-k_0\frac{N\alpha_i}{2}z} \qquad (2\text{-}83)$$

In the opposite limit of a dense, strongly resonant medium, one would find

$$n(\omega = \omega_{rk}) \sim i\sqrt{2} \qquad (2\text{-}84)$$

yielding for plane wave propagation the limit

$$E(z) = E(0)e^{-\sqrt{2}k_0 z} \qquad (2\text{-}85)$$

indicating that the wave would be totally absorbed in less than a wavelength of propagation. In general, the index as a function of angular frequency might appear as in Figure 2.21 for a medium with three resonances.

The situation in a semiconductor is in a sense a cross between the two above discussed cases. Consider the band diagram of Figure 2.22, where it is

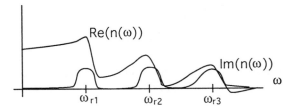

**FIGURE 2.21.**    Variation of the index of refraction with angular frequency $\omega$ for a material with three resonances.

considered that at the operating temperature, there are a number of electrons in the conduction band. Various interactions between the electrons and an electric field can occur. An applied field will tend to polarize the valence band electron clouds, much as in a dielectric. At the same time, however, an electric field will produce a conduction current given by $\sigma E$. Therefore, when one is operating near a dielectric resonance frequency, there will be contributions to $n_i$ from both conduction and valence band effects. Of course, the resonance of most interest is the optical one at angular frequency $\omega = \epsilon_g/\hbar$ where $\epsilon_g$ is the gap energy and $h = \hbar 2\pi$ is Planck's constant. Photons exciting this transition can be stimulatedly absorbed or stimulatedly amplified, or they can be absorbed by conduction band electrons in the first step of an Auger process. (An Auger process is one in which an electron absorbs a second photon before relaxation.) All of these processes can contribute to $n_i$.

Probably the main point to be gathered from the above discussion is that the index of refraction can always be expressed as a sum of a real and an imaginery part. In this sense, one can always write that

$$\mathbf{J}_{\mathrm{eff}} = \sigma_{\mathrm{eff}}(\omega)\mathbf{E}(\omega) \tag{2-86}$$

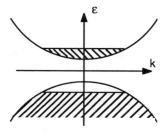

**FIGURE 2.22.**    Typical $\epsilon$-$k$ (energy-momentum) diagram for a direct gap semiconductor at a temperature high enough to cause the valence band to partially ionize, giving rise to a conduction band electron density.

where $\sigma_{\text{eff}}(\omega)$ can be defined by

$$\sigma_{\text{eff}} = \epsilon_i \omega \tag{2-87}$$

where the $\epsilon_i$ is related to $n(\omega)$ by relation (2-71) and the definition

$$\hat{\epsilon} = \epsilon_r + i\epsilon_i \tag{2-88}$$

Formally, therefore, one can solve the waveguide mode problem as was performed for real indices by just letting the various indices become complex. For example, consider the asymmetric slab problem as illustrated in Figure 2.23. One must find the same dispersion relation for this problem as was found before, namely

$$\tan \hat{\gamma}L = \frac{\hat{\gamma}(\hat{\kappa}_+ + \hat{\kappa}_-)}{\hat{\gamma}^2 - \hat{\kappa}_+ \hat{\kappa}_-} \tag{2-89}$$

except that now all of the propagation constants in all of the layers have become complex and are therefore appended with a hat to indicate their complex nature. One can write (2-89) in terms of its real and imaginary parts to obtain

$$\text{Re}(\hat{\gamma}^2 - \kappa_+\kappa_-) \tanh \gamma_i L + \text{Im} [\hat{\gamma}^2 - \hat{\kappa}_+\hat{\kappa}_-] \tan \gamma_r L$$
$$= \text{Im} [\hat{\gamma}(\hat{\kappa}_+ + \hat{\kappa}_-)] - \text{Re}(\hat{\gamma}(\hat{\kappa}_+ + \hat{\kappa}_-)) \tan \gamma_r L \tanh \gamma_i L \quad \text{(a)}$$

$$- \text{Im} (\hat{\gamma}^2 - \kappa_+\kappa_-) \tanh \gamma_i L + \text{Re}[\hat{\gamma}^2 - \hat{\kappa}_+\hat{\kappa}_-] \tan \gamma_r L$$
$$= \text{Im} [\hat{\gamma}(\hat{\kappa}_+ + \hat{\kappa}_-)] + \text{Im} (\hat{\gamma}(\hat{\kappa}_+ + \hat{\kappa}_-)) \tan \gamma_r L \tanh \gamma_i L \quad \text{(b)} \quad \text{(2-90)}$$

but this form gives us really no more insight than the other form of (2-89). Perhaps the main point to be made is that all of the constants are complex, and therefore $\beta$ will also be complex. Therefore, the form of the field will be

$$E(\mathbf{r},t) = \text{Re}[\psi(x)e^{i\beta_r z}e^{-\beta_i z}e^{i\omega t}] \tag{2-91}$$

where now it will not in general be possible to choose $\psi$ to be real, as in each

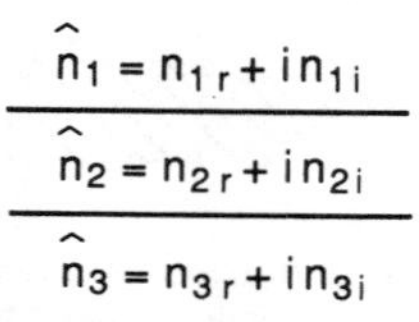

$$\hat{n}_1 = n_{1r} + in_{1i}$$
$$\hat{n}_2 = n_{2r} + in_{2i}$$
$$\hat{n}_3 = n_{3r} + in_{3i}$$

**FIGURE 2.23.**    The asymmetric slab problem with complex indices of refraction.

region the equation that it satisfies is complex. Again the condition that $A_1 = B_3 = 0$ will hold (and therefore we get the same dispersion relation as before), as the solutions to the equations in each region will be the same as before, but with the complex $\kappa$'s and $\gamma$'s. Further, the normalization condition will be the same. The problem is to find out if and when a bound mode solution exists at all.

When we considered plane polarized waves propagating in lossless slabs, we found that the equiphase planes were planes of constant $z$-value (see Figure 2.6). This was the case because we could always choose $\psi(x)$ to be real. Now we can no longer do that. In general, one must write that

$$\psi(x) = \psi_a(x)e^{i\phi(x)} \tag{2-92}$$

In this case, the constant phase planes must be solutions of the equation

$$\beta z = \omega t + \phi(x) + \text{constant} \tag{2-93}$$

Therefore, the $z$'s of a constant phase plane will vary with transverse coordinate $x$. Now the equiphase planes are the eikonal of geometrical optics. In geometrical optics, to first order at any rate, the energy propagation direction is exactly normal to the eikonal. As we saw earlier, this was indeed true for the modes of structures with real indices. The equiphases were as in Figure 2.6, and the time averaged Poynting vector of (2-55) was purely $z$-directed. Let us say that the energy flow is perpendicular to the phase front here also. Consider the three phase fronts depicted in Figure 2.24. In (a) of the figure, a plane front appears. In (b), the front is convex and the energy flow is outward. In (c), the front is concave and the energy flow is inward. Clearly, the front in (c) cannot be stable, as it will collapse. It can not possibly be a mode. The other two, however, could be.

Now, how could phase fronts like those of Figure 2.24 come about? Clearly the one in (a) could arise from propagation in a lossless medium.

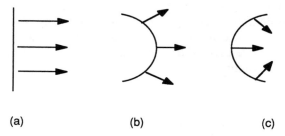

(a)　　　　　　　　(b)　　　　　　　　(c)

**FIGURE 2.24.** Three phase fronts and the associated directions of energy flow.

However, it could also arise from a situation in which the loss or gain were the same in all the layers. Now if the loss in the middle (guiding) layer became a little bit stronger than the others, the wave front would start to look like (c) and become unstable. In this sense, (a) is borderline stable. Phase fronts as in (b) could arise in two ways. If the cladding layers had higher losses than the guiding layer, there would be a net flow of energy out of the guiding layer and the phase front of (b) would result. This case is often referred to as loss guidance. In this case, the mode would damp as it propagated down the guide. The phase front in (b) could also arise from the gain being higher in the guiding layer than in the cladding. This is referred to as gain guidance and results in an electromagnetic disturbance that grows with propagation distance. Gain guidance can even serve to guide waves in structures where the real part of the guide index is smaller than the real parts of the cladding indices (index antiguiding). In fact, this was the only guiding principle in semiconductor lasers up to circa 1980. At any rate, to sum up our findings here, we can say that media with $n_{2i}$ less than both $n_{1i}$ and $n_{3i}$ can yield stable modes if $n_{2r}$ is not too much smaller than $n_{1r}$ or $n_{3r}$. No configuration can be stable if $n_{2i}$ exceeds either $n_{1i}$ or $n_{3i}$.

In many cases, one can greatly simplify the analysis based on Equations (2-90). In guiding media, one quite in general wants to choose a material in which there is not much loss. Clearly in this case $n_i \ll n_r$. In amplifying media, one would like to have the highest gain possible. Practically though, semiconductor media are the highest gain media, but still have gains that only appreciably amplify over many optical periods, requiring again that $|n_i| \ll n_r$. Now, in some cases, desirable things such as nonlinearities are greatly enhanced right in the near vicinity of strong resonances. One can use such effects to make bistable switches and other devices. These materials at these frequencies could have $|n_i| \approx n_r$, but these materials are highly dispersive and would never be used to guide light. They would be used in small places to modify light intensity, but not to take light from one place to another. Therefore, one would probably never wish to find $\beta$'s of such a material. Therefore, in analyzing guiding type structures, one can take without loss of too much generality, $|n_i| \ll n_r$. With such a simplification, one immediately notes that Equation (2-89) reduces to the usual relation (2-45). That is to say that one can solve for the $\beta$ as if the mode were lossless. One could then use the lossless $\psi(x)$ to calculate the loss factor (Harrington 1961, page 325) by the equation

$$\frac{2\beta_r\beta_i}{k_0^2} = \frac{\displaystyle\int_\infty^\infty n_i^2(x)\,\psi^2(x)\,dx}{\displaystyle\int_\infty^\infty \psi^2(x)\,dx}$$

The exponential term in $\beta_i$ in Equation (2-91) will then give a roughly proper decay factor. The only thing missing in this approximate analysis will be the actual phase curvatures. To try to calculate this, one could try to back substitute the $\beta_r$ and $\beta_i$ previously found into Maxwell's equations in each of the slab regions and recalculate the $\psi(x)$ self-consistently.

Now, in all the above discussion of gain and loss media, we have made the implicit assumption that the $n_i$ was independent of the field strength and of time. This would be a truly neat trick, but it does violate energy conservation. If we pump light into a system and it is absorbed, there will be heat generated. If this heat is not heat sinked away, it will change the properties of the material, including its refractive index which is quite usually strongly temperature dependent. But this problem is not one that we generally need to worry about, as we do not try to guide light in any material that is too strongly absorptive. However, we do have to generate light and, hopefully, amplify it when it gets weak. To do this requires a gain medium — that is, one which is in some sense inverted, that is has more electrons in excited states than ground states. As each photon that is generated must use up such an electron, in the sense that it will return the electron to its ground state, even small power amplification uses up large amounts of excited electrons. For example, a watt of near infrared (7000 Å) light, consists of about $10^{19}$ photons (that is, $1.6 \times 10^{19}$ eV/1.5 eV per photon). Therefore, a semiconductor laser with an output of 50 mW/facet needs to pump up $10^{18}$ electrons/sec to an excited state. As electrons and photons propagate at roughly the same speed (within a couple of orders of magnitude) within the lattice, this pumping process is going to have comparable time constants to the photon generation process. In this sense, the index of refraction is very much a dynamical part of the light matter interaction. Quite generally, it is a bad approximation to set an imaginary part of an index equal to a constant unless one is aware that this removes an important part of the dynamics of the problem. In this chapter, we have been concerned with finding mode structures. For this purpose, one can assume a constant index. In the next two chapters, however, we shall be concerned with dynamical effects of the light matter interaction, and for these purposes we shall need to develop a better theory. This will be done in the next chapter on the semiclassical laser equations.

## References
Born, M. and Wolf, E., 1975. *Principles of Optics,* Fifth Edition. Pergamon Press, New York.

Charczenko, W. and A.R. Mickelson, 1989. "Symmetric and Asymetric Perturbations of the Index of Refraction in Three Waveguide Optical Planar Couplers." *J. Opt. Soc. Amer.* **6,** 202–212.

Harrington, R.F., 1961. *Time Harmonic Electromagnetic Fields.* McGraw-Hill, New York.

Jackson, J.D., 1975. *Classical Electrodynamics,* Second Edition. John Wiley and
    Sons, New York.
Kurtz, C.N. and Streifer, W., 1969. "Guided Waves in Inhomogeneous Focusing
    Media. Part I: Formulation, Solution for Quadratic Inhomogeneity." *IEEE
    Trans. Microwave Theory and Tech.,* **MTT-17.**
Marcuse, D. 1982. *Optical Electronics,* Second Edition. Van Nostrand Reinhold,
    New York.
Mickelson, A.R., 1992. *Physical Optics.* Van Nostrand Reinhold, New York.

## PROBLEMS

1.  Consider a two-dimensional medium which is striated along the $x$-direction for $z > 0$ into an infinite number of slabs of width $d$ as shown in Figure 2.25. The slabs are labeled by their refractive indices $n_i$, $i$ denoting the $i^{th}$ slab. Consider a ray incident on the point $x = 0$, $z = 0$ from the region $z < 0$ with refractive index $n_0$.

    (a) Write Snell's law for the $i^{th}$ interface that the ray traverses. Note the definition of $\theta$ here.

    (b) For $d \to 0$, express $\theta_x$, the propagation angle at the coordinate $x$ in terms of the initial angle, $\theta_0$.

    (c) Assuming that $n^2(x) = n_0^2 \left[ 1 - 2\Delta \left( \dfrac{x}{a} \right)^2 \right]$, where $2\Delta \ll 1$, and that $\sin \theta_0 \leq \sqrt{2\Delta}$, show that

    $$\frac{x^2}{a^2} + \frac{\sin^2 \theta_x}{2\Delta} = \text{constant} \leq 1$$

    along a ray path. The square root of the constant is often called $R$, the mode coordinate, and can be used to label bound rays.

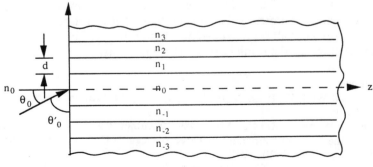

**FIGURE 2.25.**

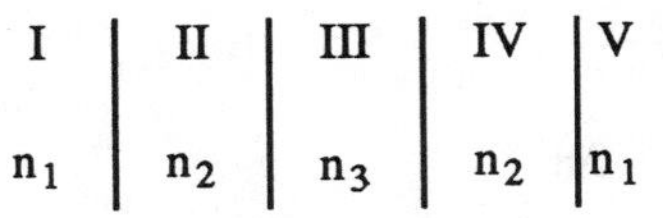

| I | II | III | IV | V |
|---|----|-----|-----|----|
| $n_1$ | $n_2$ | $n_3$ | $n_2$ | $n_1$ |

FIGURE 2.26.

    **(d)** Sketch a ray path for $R^2 \leq 1$. Label $x = a$ and $\sin \theta_0 = \sqrt{2\Delta}$ on your plot.

2. Consider a slab of width $d$ and index $n_1$ in a homogeneous infinite medium of index $n_2 > n_1$. For waves of both parallel and perpendicular polarization, find the fields in all regions as a function of $\theta_i$, the incident angle. (*Hint*: For $\theta_i < \theta_{crit}$, one can develop and sum in closed form a series of reflections.)

3. The mode cutoff condition in a metallic waveguide can be stated as $\lambda > 2d$ where d is the waveguide diameter. Using a purely ray optics argument but taking ray interference into account derive (or at least make plausible) this condition.

4. Consider a dielectric stack where $n_2 = n_1 + \Delta n$ and $\Delta n > 0$, as shown in Figure 2.26. Make a table of the *form* of the fundamental mode solution in each region when:

    **(a)** $n_2 = n_3$

    **(b)** $n_1 = n_3$

The form of the solution should contain constants which would have to be solved for from the boundary conditions.

5. Consider a dielectric stack as shown in Figure 2.27. Sketch the intensity profile of the fundamental mode solution:

    **(a)** For $\Delta n = n_1$ and $\Delta n \to 0$ where $n_1 = n_3 = n_2 - \Delta n$

    **(b)** For five equispaced values of $\Delta n$ for $0 \leq \Delta n \leq \dfrac{n_2}{2}$ where $n_1 = \dfrac{n_2}{2} + \Delta n$, $n_3 = \dfrac{n_2}{2} - \Delta n$

6. Consider a slab waveguide, where $n \sim 2$ and $\Delta n \ll n$, as shown in Figure 2.28. We want to look for modes guided in the medium with index $n$, in (a), (b) and (c).

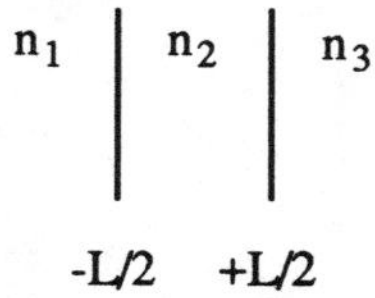

| $n_1$ | $n_2$ | $n_3$ |
|-------|-------|-------|

-L/2    +L/2

FIGURE 2.27.

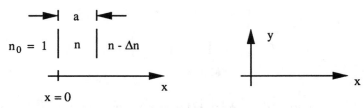

**FIGURE 2.28.**

(a) Write down the solutions assumed in each region. Find an approximate system of equations which represent the boundary conditions for the guided waves.

(b) Find the approximate TE and TM dispersion relations.

(c) Sketch the wavefunctions for the lowest order TE and TM solutions.

(d) Consider the augmented slab waveguide represented in Figure 2.29. Sketch the lowest order TE and TM wave functions.

7. Consider a weakly guiding ($n_2 = n_1 + \Delta n$ where $\Delta n \ll n_1$) symmetric slab waveguide. The TM mode dispersion relation is given by

$$\tan \gamma L = \frac{2n_1^2 n_2^2 \gamma \kappa}{n_1^4 \gamma^2 - n_2^4 \kappa^2}$$

Show that this dispersion relation is, to first order, identical to the TE mode dispersion relation. Explain why. (*Hint:* If one derives the fields to first order, it should help the thought process.)

8. A technique which employed boundary matrices to find dispersion relations is described in the text. Use this technique to:

(a) Derive the TE dispersion relation for an asymmetric slab waveguide

(b) Derive the TE dispersion relation for a periodic slab comprised of $N$ 2-layer slabs, each of thickness $t_1(t_2)$ and index $n_1(n_2)$

(c) Derive the TE dispersion relation of medium described by an arbitrary characteristic matrix $M$

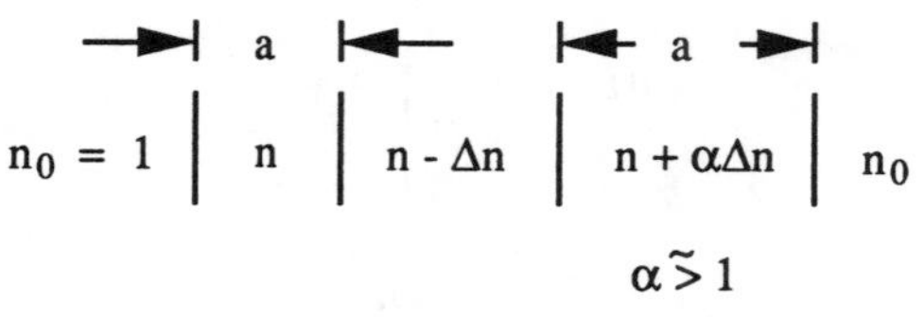

$$n_0 = 1 \quad | \quad n \quad | \quad n - \Delta n \quad | \quad n + \alpha \Delta n \quad | \quad n_0$$

$$\alpha \gtrsim 1$$

**FIGURE 2.29.**

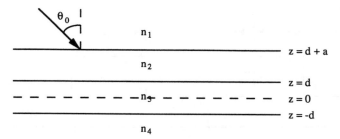

**FIGURE 2.30.**

9. Consider a two-dimensional (slab geometry) medium, where $n_1 > n_3$, $n_2 = n_0 = 1$ and $n_4 < n_3$, as depicted in Figure 2.30.
   Explain qualitatively (using simple equations and pictures) what happens when the incident angle satisfies the relation $\sin \theta_i = \dfrac{n_3}{n_1}$. For a quantitative analysis, one can refer to a paper by Midwinter in *IEEE J. Quant. Elect.* **JQE6**, 583–590 (1970).

10. Here we wish to consider a modification of the set-up of Problem 9, as depicted in Figure 2.31. How would one choose $\theta$ and $L$ to maximize the power coupled from the source to the region with index $n_3$?

11. Consider a five-layered medium, where it should be assumed that $\Delta n = n_2 - n_1 \ll n_2$, as depicted in Figure 2.32. Sketch the waveform of the fundamental modes of this structure, when:
    (a) $a_1 = a_2 \ll g$
    (b) $a_1 = g, \qquad a_2 = \tfrac{1}{2} a_1, 2a_1$
    (c) $a_1 = a_2 \gg g$
    (d) $a_1 = a_2 = g$ for several values of $\Delta n$ where $\Delta n \to 0$

12. Given that $\mathbf{B}_\omega = \mu_0 \mathbf{H}_\omega$, $\mathbf{D}_\omega = \epsilon_\omega \mathbf{E}_\omega$, and $\mathbf{J}_\omega = \sigma_\omega \mathbf{E}_\omega$, find the following in terms of $\mu_0$, $\epsilon_\omega$ and $\sigma_\omega$:

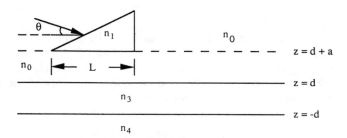

**FIGURE 2.31.**

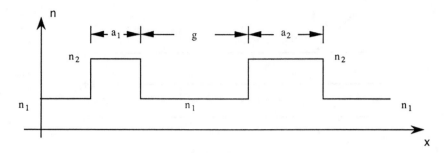

**FIGURE 3.32.**

(a) The complex permittivity $\tilde{\epsilon} = \epsilon_r + i\epsilon_i$ (*Hint:* Use $\nabla \times \mathbf{H}$ equation)
(b) The complex dielectric constant $\tilde{n}$
(c) The complex propagation constant $\tilde{k}$
(d) Sketch the evolution of the $\text{Re}[E(z)]$ for a wave propagating into such a homogeneous medium.

13. Consider the following medium

$$n_1 = n_{1r} + in_{1i}$$
$$n_2 = n_{2r} + in_{2i}$$
$$n_3 = n_{3r} + in_{3i}$$

Derive the dispersion relations (real and imaginary parts) as given by Equation (2-90). Describe with pictures and words how one might solve these equations for the medium propagation constants.

14. In the text it was discussed that one could find a perturbation theory expression for $\beta_i$ in terms of the unperturbed $\beta_r$ and $\psi(x)$. This problem concerns trying to build a perturbation series which can be used to calculate successive corrections to both the $\beta$ values and wave functions $\psi(x)$.

(a) Write down the equation satisfied by $\psi(x)$ in a medium with $n(x) = n_r(x) + in_i(x)$.

(b) Assuming that $n_i \ll n_r$, and that $\beta_i$ can be given by the perturbation expression in terms of $n_i$, $\psi^{(0)}(r)$ and $\beta_r^{(0)}$, find an equation relating $\psi^{(1)}(x)$, the first order correction to $\psi(x)$ (ie $\psi(x) = \sum_{i=1}^{\infty} \psi^{(i)}(x)$) to $\psi^{(0)}(x)$. Solve this equation in terms of the Green's function of the zeroth order operator,

$$\frac{\partial^2}{\partial x^2} + (k_0^2 n_r^2(x) - \beta_r^{(0)^2})$$

(c) How would one then find the next order correction to $\beta_r$ and $\beta_i$? Find the two equations relating $\beta_r^{(n)}$, $\beta_i^{(n)}$, $\beta_r^{n-1}$, $\beta_i^{n-1}$, and $\psi^{(n-1)}(x)$.

(d) Write out your complete set of (at least) three perturbation equations. Are there any index distributions they can be approximately solved for?

# 3

## The Semiclassical Laser Equations

In this chapter, the plan is to develop the semiclassical dynamical equations which describe time dependent light-matter interactions. The derivation presented in this chapter will assume that the medium is a simple atomic one with localized valence electrons which form a two-level system. In this sense, the derivation of this chapter will not apply per se to semiconductors, where the electrons are not localized. These restrictions, however, will be lifted in the following chapter, where attention will be directed toward semiconductor lasers.

In the second part of this chapter, following the derivation of the semiclassical equations, attention will be given to applying the equations to several simple problems, including the derivation of the rate equations, the problem of threshold and the derivation of the above threshold dynamical equation, which turns out to be the Van der Pol equation.

### 3.1 THE DERIVATION

To study the light-matter interaction, it is quite natural to begin from Maxwell's equations which we write as

$$\nabla \times \mathbf{E} = -\frac{\partial \mathbf{B}}{\partial t} \qquad \text{(a)}$$

$$\nabla \times \mathbf{H} = \mathbf{J} + \frac{\partial \mathbf{D}}{\partial t} \qquad \text{(b)} \quad \text{(3-1)}$$

The constitutive relations which are to be used to supplement (3-1) are the

**48**

usual ones

$$\mathbf{B} = \mu_0 \, \mathbf{H} \qquad \text{(a)}$$

$$\mathbf{D} = \epsilon_0 \, \mathbf{E} + \mathbf{P} \qquad \text{(b)}$$

$$\mathbf{J} = \sigma \mathbf{E} \qquad \text{(c)} \quad \text{(3-2)}$$

where the first one arises from the fact that the optical period is too short to allow collective electron motion and the third relation is only formal, in the sense that the $\sigma$ will denote all forms of loss including conduction, absorption, diffraction, and power coupling and not just conduction as (c) seems to indicate. Equation (3-2b) needs some further clarification, as it is the equation which couples the light to the matter. Indeed, the polarization $\mathbf{P}$ is the response of the material to an applied $\mathbf{E}$ field and, in that sense, is the quantity which ties us to the atomic dynamics. One can express this macroscopic polarization $\mathbf{P}$ in terms of the molecular polarizabilities $\mathbf{p}$ by the relation

$$\mathbf{P} = \sum_{\mu} \frac{\mathbf{p}_{\mu}}{V_{\mu}} \qquad (3\text{-}3)$$

where $V_{\mu}$ is the volume of the $\mu^{\text{th}}$ dipole. As the extent of an electron wave cloud is of the order of magnitude of an angstrom and the wavelength of light of the order of magnitude of a micron, one can just as well consider the dipole as pointlike and write

$$\mathbf{P} = \sum_{\mu} \mathbf{p}_{\mu} \delta(\mathbf{r} - \mathbf{r}_{\mu}) \qquad (3\text{-}4)$$

where $\mathbf{r}_{\mu}$ is the coordinate of the $\mu^{\text{th}}$ dipole. The molecular polarizability is clearly going to be given by the quantum mechanical expectation value of the dipole operator $\hat{\mathbf{p}}$, or

$$\mathbf{p}_{\mu} = \langle \hat{\mathbf{p}} \rangle_{\text{qm}} = \int \psi^*(\mathbf{r},t) \hat{\mathbf{p}} \psi(\mathbf{r},t) \, d^3\mathbf{r} \qquad (3\text{-}5)$$

where the angular brackets with the subscripted qm are the quantum mechanical expectation value which can be considered as being defined by the second equality sign in Equation (3-5), with $\psi(\mathbf{r},t)$ being the wave function of the two-level electron. But first, we wish to find the master electric field equation, and then we shall return to the necessary quantum mechanical matter equations.

Applying the curl operator to Equation (3-1a), using (3-2a), (3-1b), (3-2b), and (3-2c) in (3-1a) gives

$$\nabla \times \nabla \times \mathbf{E} = -\mu_0 \sigma \frac{\partial \mathbf{E}}{\partial t} - \mu_0 \epsilon_0 \frac{\partial^2 \mathbf{E}}{\partial t^2} - \mu_0 \frac{\partial^2 \mathbf{P}}{\partial t^2} \qquad (3\text{-}6)$$

Now, using the well-known vector identity

$$\nabla \times \nabla \times \mathbf{E} = \nabla(\nabla \cdot \mathbf{E}) - \nabla^2 \mathbf{E} \qquad (3\text{-}7)$$

one finds that

$$\nabla^2 \mathbf{E} - \mu_0 \sigma \frac{\partial \mathbf{E}}{\partial t} - \frac{1}{c^2} \frac{\partial^2 \mathbf{E}}{\partial t^2} = \mu_0 \frac{\partial^2 \mathbf{P}}{\partial t^2} + \nabla(\nabla \cdot \mathbf{E}) \qquad (3\text{-}8)$$

For the moment, let us ignore the gradient divergence term. The other four terms could be written in the form

$$\nabla^2 \begin{pmatrix} E_x \\ E_y \\ E_z \end{pmatrix} - \mu_0 \sigma \frac{\partial}{\partial t} \begin{pmatrix} E_x \\ E_y \\ E_z \end{pmatrix} - \frac{1}{c^2} \frac{\partial^2}{\partial t^2} \begin{pmatrix} E_x \\ E_y \\ E_z \end{pmatrix} = \mu_0 \frac{\partial^2}{\partial t^2} \begin{pmatrix} P_x \\ P_y \\ P_z \end{pmatrix} \qquad (3\text{-}9)$$

which shows explicitly that the equation breaks up into three scalar equations, each of the form

$$\nabla^2 A - \mu_0 \sigma \frac{\partial A}{\partial t} - \frac{1}{c^2} \frac{\partial^2 A}{\partial t^2} = \mu_0 \frac{\partial^2 B}{\partial t^2} \qquad (3\text{-}10)$$

If one writes out $\nabla(\nabla \cdot \mathbf{E})$, however, one finds that

$$\nabla(\nabla \cdot \mathbf{E}) = \frac{\partial}{\partial x}\left( \frac{\partial E_x}{\partial x} + \frac{\partial E_y}{\partial y} + \frac{\partial E_z}{\partial z} \right) \hat{\mathbf{e}}_x$$
$$+ \frac{\partial}{\partial y}\left( \frac{\partial E_x}{\partial x} + \frac{\partial E_y}{\partial y} + \frac{\partial E_z}{\partial z} \right) \hat{\mathbf{e}}_y$$
$$+ \frac{\partial}{\partial z}\left( \frac{\partial E_x}{\partial x} + \frac{\partial E_y}{\partial y} + \frac{\partial E_z}{\partial z} \right) \hat{\mathbf{e}}_z \quad (3\text{-}11)$$

which is a hopelessly coupled mess. The point here really is that the $\nabla(\nabla \cdot \mathbf{E})$ is the term that determines the wave polarization. In homogeneous, charge-free media, $\nabla \cdot \mathbf{E} = 0$, and plane wave propagation results. Plane wave propagation can be described by equations such as (3-9), where once a wave is given a polarization, it will remain polarized in that same way forever, as there is no coupling between polarization states. Here, using (3-2b),

$$\nabla(\nabla \cdot \mathbf{E}) = -\nabla \left( \frac{\nabla \cdot \mathbf{P}}{\epsilon_0} \right) \tag{3-12}$$

which gives the coupling of the polarization direction to the structure of the matter. Now, if the matter were completely homogeneous and the gradient in (3-12) were identically zero, then we could safely drop the term $\nabla(\nabla \cdot \mathbf{E})$ in (3-8). However, dropping that term means that we have assumed that the medium is perfectly homogeneous at each instant of time, and therefore we have thrown out any dynamical changes of the polarization. We will make this approximation, however, as otherwise we would not know how to proceed. Further, the resulting theory will still be able to predict linewidths, response times, etc., but not dynamical changes in polarization.

With the above argument, we see that our electric field equation is just (3-8) with the last term on the right dropped. Our problem now is to determine the $\mathbf{P}$ in (3-8) from our knowledge of atomic dynamics. The first thing that must be done is to determine what $\hat{\mathbf{p}}$ can be. We know that the classical dipole of Figure 3.1(a) has a classical dipole moment

$$\mathbf{p} = -e\mathbf{r} \tag{3-13}$$

where $\mathbf{r}$ is the separation vector. Using quantum classical correspondence, we might then conclude that

$$\hat{\mathbf{p}} = -e\hat{\mathbf{r}} \tag{3-14}$$

where $\hat{\mathbf{r}}$ is to be considered as the position operator. Now as $\mathbf{p}$ must be considered as a quantum mechanical average of $\hat{\mathbf{p}}$, we see that (3-5) can be rewritten as

$$\mathbf{p} = -e \int \psi^*(\mathbf{r},t)\hat{\mathbf{r}}\psi(\mathbf{r},t) \, d^3\mathbf{r} \tag{3-15}$$

and now all that remains to be done is to find the dynamical behavior of the wave function $\psi(\mathbf{r},t)$.

To find the behavior of $\psi(\mathbf{r},t)$ will require us to consider the Schroedinger equation

$$H(\mathbf{r},t)\psi(\mathbf{r},t) = i\hbar \frac{\partial \psi}{\partial t} \tag{3-16}$$

applied to an atom such as that depicted in Figure 3.1(b), in an electric field. To simplify the problem, we shall break up the Hamiltonian in the following manner:

$$H(\mathbf{r},t) = H_0(\mathbf{r}) + V(\mathbf{r},t) \tag{3-17}$$

where $H_0(\mathbf{r})$ part of the Hamiltonian is the part which refers to the static potential of the nucleus operating on the electron charge cloud, and the $V(\mathbf{r},t)$ refers to the dynamical interaction with the field. Unfortunately, we do not know what this potential really is for any practical atom, nor can we ever really find out. In a hydrogen atom, under the approximation that the proton is infinitely heavy with respect to the electron and ignoring the spins of both particles, one can assume that the electron moves in a uniform inverse square law field. Of course, this is wrong, as the charge distribution of the electron itself actually changes the field distribution by changing the nuclear wavefunction. The solution to this problem is to solve for the wave function in a six-dimensional rather than a three-dimensional space. Such a problem is hard enough to carry out for hydrogen but, for an $N$-electron atom, would require solving the Schroedinger equation in $3(N+1)$ dimensions, a very complicated problem indeed. Therefore, rather than worrying about detailed solutions of the problems we may not have enough available computational power to ever solve, we shall take the approach of using symmetry properties to determine the form of the solution and leave at least one parameter to be set from atomic measurements.

With the above philosophy, we shall first consider the problem of determining the properties of the stationary states of the Schroedinger equation

$$i\hbar \frac{\partial \psi(\mathbf{r},t)}{\partial t} = H_0(\mathbf{r})\psi(\mathbf{r},t) \tag{3-18}$$

for the atom, without the electromagnetic interaction. Stationary states are states whose probability density, defined by

$$P(\mathbf{r},t) = \psi^*(\mathbf{r},t)\psi(\mathbf{r},t) \tag{3-19}$$

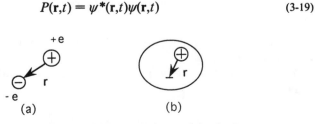

**FIGURE 3.1**  (a) Classical dipole, and (b) quantum dipole.

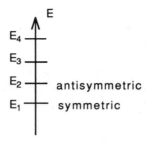

**FIGURE 3.2**    A possible diagram of the first few energy levels of a symmetric potential.

are independent of time. Such states must have time dependence of the form

$$\psi_i(\mathbf{r},t) = u_i(\mathbf{r})e^{-i\frac{E_i}{\hbar}t} \tag{3-20}$$

and therefore $u_i$ must satisfy the reduced Schroedinger equation

$$H_0 u_i(\mathbf{r}) = E_i u_i(\mathbf{r}) \tag{3-21}$$

Now under the assumption that the charge density due to the electron probability cloud does not affect the nuclear wave function and, further, that the nuclear wave function is of very small extent (ca. $10^{-15}$ m = 1 fm versus $10^{-10}$ m = 1 Å for the electron), we can assume that the potential due to the positive charge of the nucleus is a symmetric function of $\mathbf{r}$. Therefore, the states $u_i(\mathbf{r})$ of Equation (3-21) will be symmetric or antisymmetric functions of $\mathbf{r}$. Figure 3.2 illustrates what an energy level diagram might look like for the first few energy states of a system with a symmetric potential. The main point here is that the lowest energy state is symmetric and the next highest must be antisymmetric, as is illustrated in Figure 3.3. One can estimate the size and binding energy of the ground state of an atom by a reasonably simple argument. The total energy of a bound electron must be the sum of its kinetic

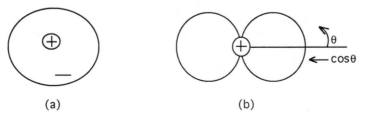

(a)                                    (b)

**FIGURE 3.3**    Illustration of two possible lowest state wave functions which appear as (a) 1s and (b) 2p hydrogenic wave functions.

energy and its potential energy. The kinetic energy must be given by $p^2/2m$, where $p$ here refers to the electron momentum and $m$ to its mass, and the potential energy must be that due to the electrostatic attraction between the positively charged nucleus and the electron wave cloud. Now as for the ground state wave function, we must have that $<p> = 0$, or else the wave function would be unbound. We see that the kinetic energy must really be due to a standard deviation of the momentum or $E_{kin} = \Delta p^2/2m$. Now, the uncertainty relation gives that

$$\Delta r \, \Delta p \geq \hbar \tag{3-22}$$

which gives one the relation for the total energy of the bound electron

$$E_{tot} = E_{kin} + E_{pot} = \frac{\hbar^2}{2m\Delta r^2} - \frac{e^2}{4\pi\epsilon_0 r} \tag{3-23}$$

Now, one can say that the $\Delta r$ in (3-23) is some measure of the wave function's extent. If there were a sharp minimum in the energy, one could perhaps, to first order at least, take $\Delta r$ to be $r$ at this minima. Consider Figure 3.4. Here it is clear that there will be a minimum of the total energy. Taking $\Delta r = r$ in (3-23), and differentiating to find the minimum $r$ yields the relations

$$r_0 = r_{min} = \frac{\hbar^2 \, (4\pi\epsilon_0)}{me^2} \tag{a}$$

$$E_{min} = \frac{-me^4}{2\hbar^2} \left( \frac{1}{4\pi\epsilon_0} \right)^2 = -\text{Rydberg} \tag{b}\ (3-24)$$

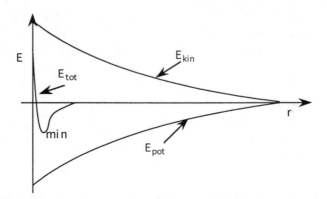

**FIGURE 3.4**    Sketch of the energies of a bound electron indicating the point of minimum total energy.

which are the Bohr radius and binding energy of the lowest state of a hydrogen atom. The argument, although rough, can also give estimates for other ground states as well as higher order states if it is generalized as in the Bohr-Sommerfeld quantization rules.

In what follows, we shall assume that we need only consider the two lowest energy states, the symmetric and antisymmetric. This assumption really entails two things. First, if we consider a system at a temperature such that without external interaction it is primarily in its ground state and, second, if we assume that $\Delta E = E_2 - E_1$ corresponds only to the 1-2 transition and to no other, then it is unlikely that, if the excitation frequency is close to $\omega = \Delta E/\hbar$, the electron can be in any other state than 1 or 2. Therefore, we consider only two states, which for generality we call $a$ and $b$, which are defined by

$$H_0(\mathbf{r})u_a(\mathbf{r}) = E_a u_a(\mathbf{r}) \tag{a}$$

$$H_0(\mathbf{r})u_b(\mathbf{r}) = E_b u_b(\mathbf{r}) \tag{b} \quad (3\text{-}25)$$

where $u_a(\mathbf{r})$ is the symmetric state and $u_b(\mathbf{r})$ is the antisymmetric state. Now as we shall consider that $H_0(\mathbf{r})$ is a "good" operator, we can assume that $u_a(\mathbf{r})$ and $u_b(\mathbf{r})$ are orthogonal in the sense that

$$\int u_a(\mathbf{r}) \, u_b(\mathbf{r}) \, d^3\mathbf{r} = 0 \tag{3-26}$$

Further, as the probability of finding a particle someplace in space is unity, or

$$\int P(\mathbf{r},t) \, d^3\mathbf{r} = \int \psi^*(\mathbf{r},t)\psi(\mathbf{r},t) \, d^3\mathbf{r} = 1 \tag{3-27}$$

we can write that

$$\int u_a^2(\mathbf{r}) \, d^3\mathbf{r} = \int u_b^2(\mathbf{r}) \, d^3\mathbf{r} = 1 \tag{3-28}$$

where we have tacitly assumed (without loss of too much generality) that the spatial part of the wave function is real. In what follows, we shall take (3-25), (3-26), and (3-28) as the basic definitions of $u_a(\mathbf{r})$ and $u_b(\mathbf{r})$.

In general, a wave function can be expressed as a sum over the time independent wave functions, of the form

$$\psi(\mathbf{r},t) = \sum_i c_i u_i(\mathbf{r}) e^{-i\frac{E_i}{\hbar}t} \tag{3-29}$$

where the $c_i$'s are excitation coefficients to be determined from some initial conditions. In our problem, we have already assumed that only two bound modes are important, and therefore we can express our wave function in the form

$$\psi(\mathbf{r},t) = c_a u_a(\mathbf{r})e^{-i\frac{E_a}{\hbar}t} + c_b u_b(\mathbf{r})e^{-i\frac{E_b}{\hbar}t} = c_a \psi_a(\mathbf{r},t) + c_b \psi_b(\mathbf{r},t) \quad (3\text{-}30)$$

It can be of interest to look at some specific examples of this wave function before continuing. Let us say, for example, that $c_a = 1$ and $c_b = 0$. The probability cloud $P(\mathbf{r},t) = \psi^*(\mathbf{r},t)\psi(\mathbf{r},t)$ about the nuclei may well appear as is depicted in Figure 3.5. The important point in both Figure 3.5(a) and (b) (for $c_a = 0$, $c_b = 1$) is that these states are time independent and, in fact, they are the only two possible states that are time independent. Now, if we note that the amplitude that the electron is in state $c$ is given by

$$A_c(\mathbf{r},t) = \int \psi_c^*(\mathbf{r},t)\psi(\mathbf{r},t) \, d^3\mathbf{r} \quad (3\text{-}31)$$

we see immediately that the probabilities of the atom being in state $a$ and state $b$ will be given by

$$P_a(\mathbf{r},t) = |C_a|^2 \quad \text{(a)}$$

$$P_b(\mathbf{r},t) = |C_b|^2 \quad \text{(b)} \quad (3\text{-}32)$$

We therefore see that the time independent states are the pure states. In general, $P(\mathbf{r},t)$ for the $\psi$ of (3-30) is given by

$$P(\mathbf{r},t) = |c_a|^2 u_a^2(\mathbf{r}) + |c_b|^2 u_b^2(\mathbf{r}) + 2\text{Re}[c_a^* c_b u_a u_b e^{-i\frac{\Delta E}{\hbar}t}] \quad (3\text{-}33)$$

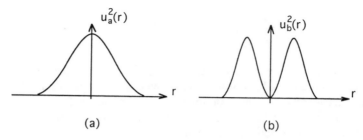

(a)                                   (b)

**FIGURE 3.5**    Depictions of the probability clouds for the two stationary states defined by (a) $c_a = 1$, $c_b = 0$ and (b) $c_a = 0$, $c_b = 1$.

The time evolution of (3-33) for $c_a = c_b = 1/\sqrt{2}$ is sketched out in Figure 3.6. The point here is that the wave function oscillates back and forth across the nuclei. This is exactly the motion we would associate with a dipole moment, and soon we shall see that this is exactly what is occurring.

Our next problem here is to find a form for the interaction potential $V(\mathbf{r},t)$ due to the radiation. Recalling that the classical energy density $E_{em}$ due to an electromagnetic field is given by

$$E_{em} = \mathbf{E} \cdot \mathbf{D} + \mathbf{B} \cdot \mathbf{H} = \epsilon_0 \, \mathbf{E} \cdot \mathbf{E} + \mu_0 \, \mathbf{H} \cdot \mathbf{H} + \mathbf{E} \cdot \mathbf{P} \qquad (3\text{-}34)$$

where the terms in $\epsilon_0$ and $\mu_0$ are the energies stored in the fields themselves, and the $\mathbf{E} \cdot \mathbf{P}$ is the interaction energy between the matter and the field. Therefore, the classical interaction energy $V(\mathbf{r},t)$ between a single dipole and an electric field could be given by

$$V(\mathbf{r},t) = -\mathbf{E} \cdot \mathbf{p} = -e\mathbf{E} \cdot \mathbf{r} \qquad (3\text{-}35)$$

where the classical relation for the microscopic dipole moment of (3-13) has been used. The minus sign in front of the $\mathbf{E} \cdot \mathbf{p}$ in (3-35) is due to the fact that it was a positive sign in (3-34). That is, the $\mathbf{E} \cdot \mathbf{p}$ energy adds to the field energy but subtracts from the electron's energy.

We now have the Schroedinger equation

$$(H_0(\mathbf{r}) - eE(\mathbf{r},t) \cdot \mathbf{r})\psi(\mathbf{r},t) = i\hbar \frac{\partial \psi}{\partial t} \qquad (3\text{-}36)$$

to solve. As our potential is now time varying, it is not immediately clear how

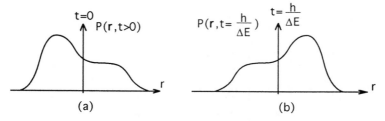

FIGURE 3.6    Sketch of the probability distribution of (3-33) for two "extreme" times (a) $t = 0$ and (b) $t = \dfrac{2\hbar\pi}{\Delta E}$.

to use our time independent solutions to find a solution here. However, as was discussed before, if our system is constrained such that it has only appreciable probability to be in two states and no others, then the time-independent wave functions of these two states approximately form a complete set of spatial states. As any disturbance in the system at any given time can be expressed as a sum over a complete set of states, one can say that, approximately, any disturbance could be described as

$$\psi(\mathbf{r},t) = C_a(t)u_a(\mathbf{r})e^{-i\frac{E_a}{\hbar}t} + C_b(t)u_b(\mathbf{r})e^{-i\frac{E_b}{\hbar}t} \tag{3-37}$$

where the coefficients $C_a(t)$ and $C_b(t)$ now contain all of the time dependence. As before, the probabilities of being in states a and b are given by

$$P_a(t) = |C_a(t)|^2 \tag{a}$$

$$P_b(t) = |C_b(t)|^2 \tag{b} \quad (3-38)$$

Now is perhaps an appropriate time to make some comments about the validity of the expansion of (3-37). Consider, as depicted in Figure 3.7, the Fourier time spectra of $\psi(\mathbf{r},t)$ for various assumptions about the variations of $C_a(t)$ and $C_b(t)$. Before some time, we can safely assume the system is in thermodynamic equilibrium and therefore, at this time, the system is in its ground state $a$ due to the fact that the thermal energy $kT$ at room temperature is very small compared to the optical transition energy $\hbar\omega_0$. Therefore, if the perturbation frequency is near the transition frequency, and the perturbation is in some sense weak, in that sense that the ground state will never be completely depopulated, then direct excitation of states $c$ and $d$ by the perturbation should be second order. Or at least, if the coefficients are constant or slowly varing, one can think about a limit where, for resonant field interaction, states $c$ and $d$ could be excluded from the interaction. For coefficients which vary rapidly with respect to the center frequencies this could not

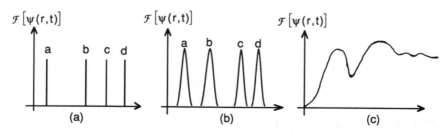

**FIGURE 3.7**    Fourier time spectra of the wave function of (3-37) for (a) constant coefficients, (b) slowly varying coefficients, and (c) rapidly varying coefficients.

be the case, as all the wave functions will be strongly coupled, as in the case of Figure 3.7(c). The point is that (3-37) in some sense is already an expansion for slowly varying coefficients. This fact will be used extensively in the derivation that follows.

Plugging the expansion of (3-37) into (3-36), one obtains

$$[H_0(\mathbf{r}) + V(\mathbf{r},t)]\, C_a(t)u_a(\mathbf{r})e^{-i\frac{E_a}{\hbar}t}$$
$$+ [H_0(\mathbf{r}) + V(\mathbf{r},t)]\, C_b(t)u_b(\mathbf{r})e^{-i\frac{E_b}{\hbar}t}$$
$$= i\hbar \frac{\partial}{\partial t}\,[C_a(t)u_a(\mathbf{r})e^{-i\frac{E_a}{\hbar}t} + C_b(t)u_b(\mathbf{r})e^{-i\frac{E_b}{\hbar}t}] \quad (3\text{-}39)$$

Now using the facts that $H_0(\mathbf{r})u_a(\mathbf{r}) = E_a u_a(\mathbf{r})$ and $H_0(\mathbf{r})u_b(\mathbf{r}) = E_b u_b(\mathbf{r})$, one finds that

$$i\hbar \frac{\partial C_a}{\partial t}\, u_a(\mathbf{r})e^{-i\frac{E_a}{\hbar}t} + i\hbar \frac{\partial C_b}{\partial t}\, u_b(\mathbf{r})e^{-i\frac{E_b}{\hbar}t}$$
$$= VC_a(t)u_a(\mathbf{r})e^{-i\frac{E_a}{\hbar}t} + VC_b(t)u_b(\mathbf{r})e^{-i\frac{E_b}{\hbar}t}] \quad (3\text{-}40)$$

Using the orthogonality relations that $\int u_a^2(\mathbf{r})\, d^3\mathbf{r} = \int u_b^2(\mathbf{r})\, d^3\mathbf{r} = 1$ and $\int u_a(\mathbf{r})u_b(\mathbf{r})\, d^3\mathbf{r} = 0$, one can show that

$$i\hbar \frac{\partial C_a}{\partial t}\, e^{-i\frac{E_a}{\hbar}t} = V_{aa}C_a(t)e^{-i\frac{E_a}{\hbar}t} + V_{ab}C_b(t)e^{-i\frac{E_b}{\hbar}t} \quad \text{(a)}$$

$$i\hbar \frac{\partial C_b}{\partial t}\, e^{-i\frac{E_b}{\hbar}t} = V_{ab}C_a(t)e^{-i\frac{E_a}{\hbar}t} + V_{bb}C_b(t)e^{-i\frac{E_b}{\hbar}t} \quad \text{(b)} \quad (3\text{-}41)$$

where

$$V_{aa} = \int u_a(\mathbf{r})V(\mathbf{r},t)u_a(\mathbf{r})\, d^3\mathbf{r} \quad \text{(a)}$$
$$V_{bb} = \int u_b(\mathbf{r})V(\mathbf{r},t)u_b(\mathbf{r})\, d^3\mathbf{r} \quad \text{(b)}$$
$$V_{ab} = V_{ba} = \int u_a(\mathbf{r})V(\mathbf{r},t)u_b(\mathbf{r})\, d^3\mathbf{r} \quad \text{(c)} \quad (3\text{-}42)$$

There are still a couple of reasonable approximations that can be made to simplify the form of (3-41). In particular, as was previously discussed, the spatial extent of the $u$'s are on the order of an angstrom (Equation (3-24a)), yet the binding energy is of the order of electron volts. Energies of electron volts correspond to wavelengths of roughly a micron, which is 10,000 times larger than an angstrom. The situation looks like that depicted in Figure 3.8. As the field cannot curve its phase front over a distance smaller than a

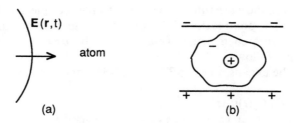

**FIGURE 3.8**    Phase curvature of (a) an incident disturbance relative to the extent of an atom and (b) a model in which the electric field can be spatially replaced by a parallel plate capacitor.

wavelength, the wave looks essentially plane to the atom. In a major sense, therefore, the field's spatial variation can be considered to be constant, as is the field in a parallel plate capacitor, but with an optical time variation. Therefore one can approximate

$$V(\mathbf{r},t) = -e\mathbf{E}(t) \cdot \mathbf{r} \tag{3-43}$$

But the potential of (3-43) is now a purely odd function of $\mathbf{r}$, and therefore $V_{aa}$ and $V_{bb}$ become equal to zero from symmetry, and the system of (3-41) can be reduced to

$$\frac{dC_a}{dt} = \frac{1}{i\hbar} C_b(t)\mathbf{E}(t) \cdot \mathscr{P}e^{-i\omega t} \tag{a}$$

$$\frac{dC_b}{dt} = \frac{1}{i\hbar} C_a(t)\mathbf{E}(t) \cdot \mathscr{P}e^{i\omega t} \tag{b} \quad (3\text{-}44)$$

where the identifications

$$\mathscr{P} = e \int u_a(\mathbf{r})\mathbf{r}u_b(\mathbf{r}) \, d^3\mathbf{r} \tag{a}$$

$$\omega = \frac{E_b - E_a}{\hbar} \tag{b} \quad (3\text{-}45)$$

have been made.

One could take Equation (3-45) together with the master field equation

$$\nabla^2\mathbf{E} - \mu_0\sigma \frac{\partial\mathbf{E}}{\partial t} - \frac{1}{c^2}\frac{\partial^2\mathbf{E}}{\partial t^2} = \mu_0 \frac{\partial^2\mathbf{P}}{\partial t^2} \tag{3-46}$$

as the semiclassical equations. As we will see presently, there are several reasons for not doing this. At any rate, rather than needing the $C$'s to complement (3-46), one wants to know what the $\mathbf{p}_\mu$'s are as defined by equations (3-4) and (3-5). Taking the quantum mechanical dipole operator to be as in Equation (3-14) and inserting between $\psi^*(\mathbf{r},t)$ and $\psi(\mathbf{r},t)$, where $\psi(\mathbf{r},t)$ is as in (3-37), one can write that

$$\mathbf{p}_\mu = -e \int (C_a^*(t)u_a(\mathbf{r})e^{i\frac{E_a}{\hbar}t} + C_b^*(t)u_b(\mathbf{r})e^{i\frac{E_b}{\hbar}t})\mathbf{r}$$

$$(C_a(t)u_a(\mathbf{r})e^{-i\frac{E_a}{\hbar}t} + C_b(t)u_b(\mathbf{r})e^{-i\frac{E_b}{\hbar}t})\, d^3\mathbf{r}$$

$$= C_a^*(t)C_e(t)e^{-i\omega t}\mathscr{P} + C_a(t)C_b^*(t)e^{-i\omega t}\mathscr{P}$$

$$= \alpha_\mu(t)\mathscr{P} + \alpha_\mu^*(t)\mathscr{P} \tag{3-47}$$

where the last equality sign can serve to define the complex dipole moment coefficient $\alpha_\mu(t)$. Note that $\alpha_\mu(t)$ is a complex scalar but can be related to the vector quantities

$$\alpha_\mu(t)\mathscr{P} = \mathbf{p}_\mu^{(+)} \tag{a}$$

$$\alpha_\mu^*(t)\mathscr{P} = \mathbf{p}_\mu^{(-)} \tag{b} \quad (3\text{-}48)$$

where the $\mathbf{p}_\mu^{(+)}$ ($\mathbf{p}_\mu^{(-)}$) is the positive (negative) frequency contribution to the complex dipole moment. Now, brutally plugging in the definition for $\alpha_\mu(t)$ in terms of $C_a$, $C_b$, and $\omega$, one can find a dynamical equation for $\alpha_\mu(t)$ as follows:

$$\frac{d\alpha_\mu}{dt} = \frac{dC_a^*(t)}{dt} C_b(t)e^{-i\omega t} + C_a^*(t)\frac{dC_b(t)}{dt} e^{-i\omega t}$$

$$- i\omega C_a^*(t)C_b(t)e^{-i\omega t} \tag{3-49}$$

Substituting (3-44) into (3-49), one can obtain

$$\frac{d\alpha_\mu}{dt} = -i\omega\alpha_\mu(t) - \frac{1}{i\hbar} \mathbf{E}(t) \cdot \mathscr{P} \, d_\mu \tag{3-50}$$

where the inversion parameter $d_\mu$ is defined by

$$d_\mu = |C_b(t)|^2 - |C_a(t)|^2 \tag{3-51}$$

As is evident from its definition, $d_\mu$ varies between $-1$ and $+1$ with $-1$ corresponding to a pure ground state wave function and $+1$ corresponding to a pure excited state wave function. One can derive the dynamical equation for $d_\mu$, in an analogous manner to the one calculated in (3-49) and (3-50) for $\alpha_\mu$, to obtain

$$\frac{dd_\mu}{dt} = \frac{2}{i\hbar} \, \mathbf{E}(t) \cdot [\alpha_\mu^*(t)\mathscr{P} - \alpha_\mu(t)\mathscr{P}] \tag{3-52}$$

Now the system of equations of (3-46)–(3-48), (3-50), and (3-52) is a complete set of equations for the coupled atomic state/radiation field. The system still is not quite complete, however. For example, although there is a damping term in (3-46), both (3-50) and (3-52) are totally undamped. This could lead to a situation in which a field which was turned on and off in the past could lead to a continuing $\alpha_\mu$ far into the future. One would expect such permanent material oscillations to be damped out. Indeed, the problem here stems from the fact that our closed quantum mechanical equations were loss free, as closed quantum mechanical equations must be from their Hamiltonian nature. Therefore, we need to put in phenomenological damping by hand. Let us first consider Equation (3-52) for $d_\mu$. For a two-level system to come to equilibrium, there must be an equilibrium between stimulated and relaxation processes as depicted in Figure 3.9. For simplicity, one could consider a case in which the optical absorption of the medium were high and therefore the relaxation processes were the important ones. In such a case, the right-hand side of (3-52) goes to zero, which cannot quite be correct. Relaxation processes in gases and liquids would generally be due to collisions, whereas in a solid they would be due to phonon scattering. Now, whether the relaxation process is due to particle or phonon scatterings at a given temperature, there will be a characteristic time for an excited particle to be thermalized. Let us call this time $\tau_d$. Further, at a given temperature greater than zero, $d_\mu$ will have a value greater than $-1$, as scatterings can have the effect of exciting atoms to their upper state. Let us call this equilibrium value of $d_\mu$, $d_0$. If one were to prepare the $\mu^{\text{th}}$ atom in a state defined by $d_{\mu i}$, this state would have to decay to the state $d_0$ in a time $\tau_d$. Such behavior can

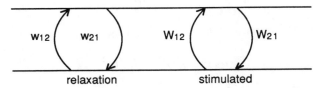

**FIGURE 3.9**    Competition in a two-level system between stimulated and relaxation processes.

be described by a differential equation of the form

$$\frac{dd_\mu}{dt} = \frac{d_0 - d_\mu}{\tau_d} \qquad \text{(a)}$$

$$d_\mu(0) = d_{\mu i} \qquad \text{(b)} \quad \text{(3-53)}$$

with solutions

$$d_\mu = d_{\mu i} e^{-t/\tau_d} + d_0(1 - e^{-t/\tau_d}) \qquad \text{(3-54)}$$

Another fix-up of (3-53) which we shall have occasion to use later has to do with the definition of the $d_0$, although we said that the reason for $d_0$ to be different from $-1$ was thermal in origin. Certainly the $\tau d$ is of thermal origin, but the $d_0$ is not necessarily. If it were, one could never achieve laser action, as one can never thermally invert a medium. A medium can be pumped by the scheme depicted in Figure 3.10. The idea is to pump a transition between two nonlasing states. The upper state should be a short-lived one that decays nonradiatively to a metastable state $b$. This pumping will essentially cause a quasi-equilibrium which can elevate the value of $d_0$ to a positive value (inversion). Because the situation is that of quasi-equilibrium, the thermal value of $\tau_d$ could still be used. We therefore can now assume that our full $d_\mu$ equation can be written in the form

$$\frac{dd_\mu}{dt} = \frac{d_0 - d_\mu}{\tau_d} + \frac{2}{i\hbar} \, \mathbf{E}(t) \cdot [\alpha_\mu^*(t)\mathscr{P} - \alpha_\mu(t)\mathscr{P}] \qquad \text{(3-55)}$$

We next turn to Equation (3-50) for the complex polarization coefficient. The situation with this equation is analogous to what we saw with (3-52), although the constant in the fix-up will be of a different order of magnitude than that in (3-55). The point here is that scattering processes will also tend to cause a decay of the complex polarization coefficient. This decay will be much more rapid than that of the inversion for a reason depicted in Figure 3.11. The point is that it is more likely that a phonon elastically scatters from

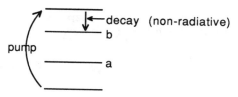

**FIGURE 3.10**    Four-level laser system.

FIGURE 3.11    (a) Phonon/atom interaction in which the inversion value $d_\mu$ is lowered from $+1$ to $-1$ and (b) phonon/atom interaction in which the $\alpha_\mu$ value is driven to zero without any attendant change in the inversion value.

an atom as in Figure 3.10(b) than both the phonon and atom change state during the process. In order for the inversion to be changed, the inelastic process must occur. However, the elastic process could well change the phase between the $C_a$ and $C_b$ of Equation (3-47), thereby driving the $\alpha_\mu$ to zero without affecting the relative values of $C_a$ and $C_b$ or, for that matter, the inversion $d_\mu$. For this reason, a term $-\gamma \alpha_\mu$ with the value of $\gamma = 1/\tau_a$ should be included on the right-hand side of Equation (3-50) to yield

$$\frac{d\alpha_\mu}{dt} = (-i\omega - \gamma)\alpha_\mu - \frac{1}{i\hbar}\mathbf{E}(t) \cdot \mathscr{P}d_\mu \qquad (3\text{-}56)$$

Before looking into some of the properties of lasers, it can be of interest to derive the macroscopic form of these equations, which are so useful in the study of pulse propagation through active media. As was mentioned already in Equation (3-4) of this chapter, macroscopic quantities can be defined as delta function weighted sums over corresponding microscopic quantities. Making the definitions

$$\mathbf{P} = \mathbf{P}^{(+)} + \mathbf{P}^{(-)} = \sum_\mu \langle [\mathbf{p}_\mu^{(+)}\delta(\mathbf{r} - \mathbf{r}_\mu) + \mathbf{p}_\mu^{(-)}\delta(\mathbf{r} - \mathbf{r}_\mu)] \rangle_{cl} \quad \text{(a)}$$

$$D = \left\langle \sum_\mu d_\mu \delta(\mathbf{r} - \mathbf{r}_\mu) \right\rangle_{cl} \qquad \text{(b)} \quad (3\text{-}57)$$

one can immediately obtain from (3-46)–(3-48), (3-55), and (3-56) the system

$$\nabla^2\mathbf{E} - \frac{1}{c^2}\frac{\partial^2\mathbf{E}}{\partial t^2} - \mu_0\sigma\frac{\partial\mathbf{E}}{\partial t} = \mu_0\left(\frac{\partial^2\mathbf{P}^{(+)}}{\partial t^2} + \frac{\partial^2\mathbf{P}^{(-)}}{\partial t^2}\right) \qquad \text{(a)}$$

$$\frac{d\mathbf{P}^{(+)}}{dt} = (-i\omega - \gamma)\mathbf{P}^{(+)} + \frac{1}{i\hbar}(\mathbf{E} \cdot \mathscr{P})\mathscr{P}D \qquad \text{(b)}$$

$$\frac{dD}{dt} = \frac{D_0 - D}{\tau_d} - \frac{2}{i\hbar}\mathbf{E} \cdot (\mathbf{P}^{(-)} - \mathbf{P}^{(+)}) \qquad \text{(c)} \quad (3\text{-}58)$$

which, when the conjugate of (3-58b) is included, is a closed system of equations for the macroscopic **E** field, the complex polarizability, and the macroscopic inversion. This is precisely the system of equations that can be used to treat such effects as self-induced transparency and solitary wave propagation in active media. But these are topics for another book

To understand laser operation, one must consider the behavior of a Maxwell-Bloch-type system inside a cavity resonator. The simplest cavity resonator that comes to mind is an empty Fabry-Perot resonator, as is depicted in Figure 3.12. First let us consider what happens when the reflectivities are allowed to become negligible. In this case, a field in the cavity can always be expressed as a plane wave of the form

$$\mathbf{E}(\mathbf{r},t) = \text{Re}[Ae^{i\frac{\omega}{c}z}e^{-i\omega t}] \tag{3-59}$$

where there is no restriction on what values $\omega$ may take on. If one considers the converse case where the $R$'s are allowed to approach unity, then the situation changes, as in this case only certain $\omega$'s can exist in the cavity. Were one to denote the cavity length to be $L$, then the angular frequencies $\omega_N$ for which a standing wave pattern with zeroes at the mirrors, defined by

$$\frac{\omega_N}{c}L = \pi N \tag{3-60}$$

would be the only set of frequencies allowed inside the cavity. One could then expand any (nonmonochromatic) field existing in the cavity in the form

$$\mathbf{E}(\mathbf{r},t) = \sum (\mathbf{A}_N \cos \omega_N t + \mathbf{B}_N \sin \omega_N t) \sin \frac{\omega_N}{c}Z \tag{3-61}$$

which can be expanded in the form

$$\mathbf{E}(\mathbf{r},t) = \sum E_N(t)\mathbf{u}_N(z) = \sum_\lambda E_\lambda(t)\mathbf{u}_\lambda(Z) \tag{3-62}$$

where the $\lambda$'s are a counting variable related to the $\omega_N$'s.

**FIGURE 3.12**    Fabry-Perot resonator with mirrors of reflectivity $R$.

The situation is slightly more involved although formally the same in more complicated guides. In a guide which was an $x$ and $y$ varying index of refraction, one must solve a boundary value problem to obtain a dispersion relation, which gives the $\beta_i$'s of the modal expression

$$\mathbf{E}_i(\mathbf{r},t) = \text{Re}[\psi(x,y)e^{i\beta_i(\omega)z}e^{-i\omega t}] \qquad (3\text{-}63)$$

where any arbitrary, monochromatic disturbance in the cavity can be expanded in the form

$$\mathbf{E}_{\text{tot}}(\mathbf{r},t) = \sum_i a_i \mathbf{E}_i(\mathbf{r},t) \qquad (3\text{-}64)$$

If the medium is placed between perfectly reflecting mirrors at $z = 0$ and $z = L$, again, as in the plane wave case, only certain $\omega$'s will be able to exist in the cavity. These $\omega$'s must be solutions of the equation

$$\beta_i(\omega_N)L = N\pi \qquad (3\text{-}65)$$

which is the required generalization of Equation (3-60). Clearly then, one can relate the $\omega_N$'s to a counting variable $\lambda$ to enable one to write the nonmonochromatic mode field expansion in the cavity as

$$\mathbf{E}_{\text{tot}}(\mathbf{r},t) = \sum_\lambda E_\lambda(t)\mathbf{u}_\lambda(\mathbf{r}) \qquad (3\text{-}66)$$

where the $\mathbf{u}_\lambda(\mathbf{r})$ must be given by

$$\mathbf{u}_\lambda(\mathbf{r}) = \psi_\lambda(x,y) \sin \beta_\lambda(\omega_\lambda)z \qquad (3\text{-}67)$$

Now, one wants to use the expansion of (3-66) to simplify the electric field Equation (3-46) and, in particular, to remove the spatial coordinate variation of the equation and map it onto the $\lambda$ basis developed above. Clearly, from separation of variables, one can write that $\mathbf{u}_\lambda(\mathbf{r})$ must satisfy

$$\nabla^2\mathbf{u}_\lambda(\mathbf{r}) + k_\lambda^2\mathbf{u}_\lambda(\mathbf{r}) = 0 \qquad (3\text{-}68)$$

subject to boundary conditions on the mirrors at 0 and $L$. As the operator in (3-68) is a nice one, the $\mathbf{u}_\lambda(\mathbf{r})$ must be orthogonal, and one can choose them to be orthonormal such that

$$\int \mathbf{u}_\lambda(\mathbf{r}) \cdot \mathbf{u}_{\lambda'}(\mathbf{r}) \, d^3\mathbf{r} = \delta_{\lambda\lambda'} \qquad (3\text{-}69)$$

Recall that, in Chapter 2, when we normalized propagating solutions, we used a power normalization. The point here is that the field in the cavity is not propagating but instead is standing, and further, we have yet to normalize the $E_\lambda(t)$ coefficients, so (3-69) does not actually fix the normalization. We shall see that when we do normalize the $E_\lambda(t)$, we fix the normalization to a so-called *photon normalization*.

Plugging the expansion of (3-66) into Equation (3-46), one finds that

$$\nabla^2 \sum_\lambda \mathbf{E}_\lambda(t)\mathbf{u}_\lambda(\mathbf{r}) - \mu_0\sigma \frac{\partial}{\partial t} \sum_\lambda \mathbf{E}_\lambda(t)\mathbf{u}_\lambda(\mathbf{r})$$

$$-\frac{1}{c^2}\frac{\partial^2}{\partial t^2}\sum_\lambda \mathbf{E}_\lambda(t)\mathbf{u}_\lambda(\mathbf{r}) = \mu_0 \frac{\partial^2\mathbf{P}(\mathbf{r},t)}{\partial t^2} \qquad (3\text{-}70)$$

Using (3-68) to simplify the $\nabla^2$ term in (3-70) and then multiplying all terms through by $\mathbf{u}_{\lambda'}(\mathbf{r})$ and integrating over all space such that the relations of (3-69) can be used, one obtains the relation

$$\frac{\partial^2 E_\lambda(t)}{\partial t^2} + \frac{1}{\epsilon_0}\sum_{\lambda'}\sigma_{\lambda\lambda'}\frac{\partial E_{\lambda'}(t)}{\partial t} + \omega_\lambda^2 E_\lambda(t) = -\frac{1}{\epsilon_0}\frac{\partial^2 P_\lambda(t)}{\partial t^2} \qquad (3\text{-}71)$$

where $\sigma_{\lambda\lambda'}$, and $P_\lambda(\mathbf{r},t)$ are defined by

$$\sigma_{\lambda\lambda'} = \int \sigma(\mathbf{r})\mathbf{u}_\lambda(\mathbf{r}) \cdot \mathbf{u}_{\lambda'}(\mathbf{r})\, d^3\mathbf{r} \qquad \text{(a)}$$

$$P_\lambda(t) = \int \mathbf{u}_\lambda(\mathbf{r}) \cdot \mathbf{P}(\mathbf{r},t)\, d^3\mathbf{r} \qquad \text{(b)} \quad (3\text{-}72)$$

Although it is clear that $\sigma$, any way we define it, must have spatial variation, if only because some power must leak out through the mirrors, we choose to ignore this and let the $\sigma$ be uniformly distributed throughout the cavity. This can be justified really for two reasons. One follows from the earlier discussion (Chapter 2) of the use of $\sigma$ and the nebulousness of its meaning. Why try to be precise about something we can not even precisely define? The second reason is the inconsistency of assigning loss to the mirrors after, in the derivation of the expansion in (3-66), we repeatedly assumed that they were perfectly reflecting. It is probably just as well to distribute this loss throughout the cavity than to assign it to the walls. A real solution to this problem would require one to find the modes of a universe containing a cavity with partially reflecting mirrors, a nontrivial problem at best and, further, an unsolved one. At any rate, with the constant $\sigma$ assumption, Equation (3-71) reduces to

$$\frac{\partial^2 E_\lambda(t)}{\partial t^2} + \frac{\sigma}{\epsilon_0}\frac{\partial E_\lambda(t)}{\partial t} + \omega_\lambda^2\, E_\lambda(t) = -\frac{1}{\epsilon_0}\frac{\partial^2 P_\lambda(t)}{\partial t^2} \tag{3-73}$$

The system represented by (3-73), (3-56), and (3-55) can be further simplified by using a rotating wave representation and assuming slowly varying coefficients, an assumption that was already implicit in our limiting our consideration to a two-level system. The rotating wave representation was already introduced in (3-48), where the positive and negative frequency components of **p** were split up. Of course, in saying that $\alpha(t)\mathscr{P}$ represents the positive frequency component of **p**, one has already assumed that $C_a(t)$ and $C_b(t)$ are slowly varying compared to $e^{-i\omega t}$. In the same vein, one can write that

$$E_\lambda^{(+)}(t) = A_\lambda(t)e^{-i\omega_\lambda t} \tag{a}$$

$$E_\lambda^{(-)}(t) = A_\lambda^*(t)e^{-i\omega_\lambda t} \tag{b} \quad (3\text{-}74)$$

where the $A_\lambda(t)$ is a complex constant and where, again, identification of (+) with positive frequency and the (−) with negative frequency requires that the $A_\lambda(t)$ be slowly varying with respect to $e^{-i\omega_\lambda t}$. To perform the desired simplifications, it will also be necessary to write the $P_\lambda(t)$ in terms of positive and negative frequency components. To carry this out, one can first write

$$P_\lambda^{(+)}(\mathbf{r},t) = \left\langle \sum_\mu \mathbf{p}_\mu^{(+)} \cdot \mathbf{u}_\lambda(\mathbf{r} - \mathbf{r}_\mu) \right\rangle \tag{a}$$

$$P_\lambda^{(-)}(\mathbf{r},t) = \left\langle \sum_\mu \mathbf{p}_\mu^{(-)} \cdot \mathbf{u}_\lambda(\mathbf{r} - \mathbf{r}_\mu) \right\rangle \tag{b} \quad (3\text{-}75)$$

in light of the identifications of (3-48). Using (3-48) together with (3-75) and (3-72b), one can write that

$$P_\lambda^{(+)}(t) = \sum_\mu \alpha_\mu(t)\mathbf{u}_\lambda(\mathbf{r}_\mu) \cdot \mathscr{P} \tag{a}$$

$$P_\lambda^{(-)}(t) = \sum_\mu \alpha_\mu^*(t)\mathbf{u}_\lambda(\mathbf{r}_\mu) \cdot \mathscr{P} \tag{b} \quad (3\text{-}76)$$

where the identifications of (3-48), where $\alpha(t)$ is associated with positive frequencies and $\alpha^*(t)$ is identified with negative frequencies, for slowly varying $C_1(t)$ and $C_2(t)$ at least, identifies $P_\lambda^{(+)}(t)$ with positive frequencies and $P_\lambda^{(-)}(t)$ with negative frequencies. With the identifications of (3-74) and (3-76), it is clear that one can rewrite (3-73) as a positive frequency component equation and a negative frequency component equation. The positive equation can be written as

$$\frac{\partial E_\lambda^{(+)}(t)}{\partial t} + i\omega_\lambda E_\lambda^{(+)}(t) + \frac{\sigma}{2\epsilon_0} E_\lambda^{(+)}(t) = \frac{-i\omega_\lambda}{2\epsilon_0} P_\lambda^{(+)}(t) \tag{3-77}$$

where the corollaries to the slowly varying approximation

$$\left| \frac{\partial A_\lambda}{\partial t} \right| \ll |\omega_\lambda A_\lambda| \tag{a}$$

$$\frac{\partial P_\lambda^{(+)}(t)}{\partial t} = i\omega P_\lambda^{(+)}(t) \tag{b} \quad \text{(3-78)}$$

have been used. Making the identification that $\kappa_\lambda = \sigma/2\epsilon_0$ and expanding out the $P_\lambda^{(+)}(t)$, one can write (3-77) in the form

$$\frac{\partial E_\lambda^{(+)}(t)}{\partial t} = (-i\omega_\lambda - \kappa_\lambda)E_\lambda^{(+)}(t) - \sum_\mu \frac{i\omega_\lambda}{2\epsilon_0\hbar} \mathscr{P} \cdot \mathbf{u}_\lambda(\mathbf{r}_\mu)\alpha_\mu(t) \tag{3-79}$$

A comparable (approximate) transformation can be performed on (3-56) to obtain two equations. The positive frequency part of this equation becomes

$$\frac{\partial \alpha_\mu}{\partial t} = (-i\omega - \gamma)\alpha_\mu - \frac{1}{i\hbar} \sum_\lambda E_\lambda^{(+)}(t)\mathscr{P} \cdot \mathbf{u}_\lambda(\mathbf{r}_\mu) \, d_\mu \tag{3-80}$$

As the slowly varying approximation has already been made repeatedly and almost everywhere, it stands to reason that $d_\mu$ must be centered around zero frequency. This further means that in (3-55) only products of positive frequency terms with negative frequency terms can show up. With this in mind, it can be shown that the $d_\mu(t)$ equation can be reduced to the form

$$\frac{\partial d_\mu(t)}{\partial t} = \frac{d_0(t) - d_\mu(t)}{\tau_d} + \frac{2}{i\hbar} \sum_\lambda$$
$$(E_\lambda^{(+)}(t)\mathscr{P} \cdot \mathbf{u}_\lambda(\mathbf{r}_\mu)\alpha_\mu^* - E_\lambda^{(-)}(t)\mathscr{P} \cdot \mathbf{u}_\lambda(\mathbf{r}_\mu)\alpha_\mu(t)) \tag{3-81}$$

We wish to make only one more set of transformations to get (3-79)(3-81) into the semiclassical form of the laser equations. First, we wish to define a coupling constant $g_{\mu\lambda}$ by

$$g_{\mu\lambda} = i\mathscr{P} \cdot \mathbf{u}_\lambda(\mathbf{r}_\mu) \sqrt{\frac{\omega_\lambda}{2\epsilon_0\hbar}} \tag{3-82}$$

Further, we wish to define the dimensionless electric field coefficients $b_\lambda(t)$ by

$$E_\lambda^{(+)} = i \sqrt{\frac{\hbar\omega_\lambda}{2\epsilon_0}} \, b_\lambda \tag{a}$$

$$E_\lambda^{(-)} = -i \sqrt{\frac{\hbar\omega_\lambda}{2\epsilon_0}} \, b_\lambda^* \tag{b} \quad (3\text{-}83)$$

The normalization in (3-83) is a natural one to use for the following reason. The energy $E_{em}$ per unit volume $V$ for an electromagnetic plane wave is given by

$$\frac{E_{em}}{V} = \epsilon_0 \mathbf{E} \cdot \mathbf{E}^* + \mu_0 \mathbf{H}^* \cdot \mathbf{H} = 2\epsilon_0 \mathbf{E} \cdot \mathbf{E}^* \tag{3-84}$$

Evaluating (3-84) using $\mathbf{E} = E_\lambda^{(-)}(t)\mathbf{u}_\mu(\mathbf{r})$, one finds immediately that

$$E_{em} = 2\epsilon_0 E_\lambda^{(+)} E_\lambda^{(-)} = \hbar\omega_\lambda b_\lambda^* b_\lambda = \hbar\omega_\lambda n_\lambda \tag{3-85}$$

where (3-85) can serve as a definition of $n_\lambda = b_\lambda^* b_\lambda$. One knows that the energy of a photon of wavelength $\lambda$ is just $\hbar\omega_\lambda$. Equation (3-85) states that the electromagnetic energy of wavelength $\lambda$ in a cavity is given by $n_\lambda \hbar\omega_\lambda$, which implies that $n_\lambda$ must be the photon number and $b_\lambda$ the photon normalized electromagnetic field. Using the identifications of (3-82) and (3-83) in (3-79)–(3-81), one obtains the system of equations

$$\frac{\partial b_\lambda(t)}{\partial t} = (-i\omega_\lambda - \kappa_\lambda)b_\lambda(t) - i \sum_\mu g_{\mu\lambda}^* \alpha_\mu(t) \tag{a}$$

$$\frac{\partial \alpha_\mu(t)}{\partial t} = (-i\omega - \gamma)\alpha_\mu(t) + i \sum_\lambda g_{\mu\lambda} b_\lambda(t) d_\mu(t) \tag{b}$$

$$\frac{\partial d_\mu(t)}{\partial t} = \frac{d_0(t) - d_\mu(t)}{\tau_d} + 2i \sum_\lambda (g_{\mu\lambda}^* b_\lambda^*(t)\alpha_\mu(t) - g_{\mu\lambda} b_\lambda(t)\alpha_\mu^*(t)) \tag{c} \quad (3\text{-}86)$$

The meanings of the various terms in (3-86) are reasonably clear. The first terms on the right-hand sides of (3-86a) and (3-86b) indicate that the field and polarization will exhibit decayed oscillation after an excitation. The first term on the right-hand side of (3-86c) is the term that says that, in an unpumped system, the inversion will relax back to an equilibrium state. The coupling terms, the second terms on the right-hand sides of (3-86), are slightly more complex to interpret. The easiest one to interpret is that in (3-86a). That term just indicates that energy stored in the oscillating dipole moments of the atoms is returned to the field through this driving term. As far as the coupling in (3-86b) goes, it is possible to convince oneself that this term represents the effect of stimulated processes on the polarization. If $d_\mu$ is negative, absorption processes tend to increase $\alpha_\mu$ up until $d_\mu$ goes to zero and $C_a$ becomes equal to $C_b$. For positive $d_\mu$, it is the stimulated emission processes that drive $\alpha_\mu$. Since it is the phase of $b_\lambda$ relative to $\alpha_\mu$ that determines whether absorption or emission takes place, it is clear that the coupling term properly tracks the sign of $d_\mu$. This phasing effect is even more clear in the coupling term of (3-86c). If one assumes that $g_{\mu\lambda}$ is real, then this term clearly goes to zero when $b_\lambda$ and $\alpha_\mu$ are in phase. That $g_{\mu\lambda}$ is actually imaginary simply indicates that there is a delay in the interaction of $b_\lambda$ and $\alpha_\mu$. The situation is as depicted in Figure 3.13. One can think of the field as being an external driving force for the $\alpha_\mu$ which can be represented as a mass on a spring. For a real $g_{\mu\lambda}$, the field would drive the spring directly, as in (a). For an imaginary $g_{\mu\lambda}$, the field's effect on the spring is delayed by some portion of a period. Therefore, without loss of generality, one can consider $g_{\mu\lambda}$ real for this discussion at least. The idea that the coupling goes to zero for $\alpha_\mu$ and $b_\lambda$ in phase can be understood from 3.13(a). If the field and spring are exactly in step, no energy will be exchanged. If the field leads to the spring, the spring will be forced to catch up and therefore absorb energy from the field. A lagging field will tend to absorb energy from the springs. Such are the effects of phase on field dynamics.

## 3.2 Rate Equations

It now seems time to study some of the salient characteristics of the semiclassical system. It certainly is of interest to investigate the manner in which these equations include the standard laser rate equations. To derive these

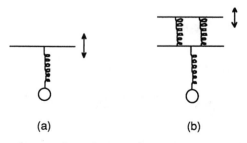

(a)      (b)

**FIGURE 3.13**    Schematic depiction of the situation in which (a) $b$ and $\alpha$ are in phase and (b) where a complex $g_{\mu\lambda}$ causes them to oscillate at a given phase angle.

equations, we first take

$$b_\lambda = B_\lambda e^{-i\omega_\lambda t} \tag{3-87}$$

where the $B_\lambda$ should be slowly varying. Next we wish to eliminate the polarization $\alpha_\mu$ adiabatically. What this means is that, if the polarization closely follows the behavior of the field, its dynamics will be identical to the field dynamics and it can therefore be eliminated from the system of equations. We already know that the $\gamma$ in the $\alpha_\mu$ equation must be much larger than $1/\tau_d$. We will further assume that it is much larger than $\kappa_\lambda$ for the $\lambda$'s of interest. As $1/\gamma = \tau_\alpha$ is the time constant for changes in the polarizability and $\gamma$ is the largest of the damping constants, then it stands to reason that $\alpha_\mu$ adjusts itself to changes in $d_\mu$ and $b_\lambda$ much faster than $d_\mu$ and $b_\lambda$ can adjust to changes in $\alpha_\mu$. Assuming a roughly constant pump therefore implies that $\alpha_\mu$ will adiabatically follow the field, and one can therefore take that

$$\frac{d\alpha_\mu(t)}{\partial t} = -i\omega_\lambda \alpha_\mu \tag{3-88}$$

Using (3-88) and (3-86), one immediately finds that

$$\alpha_\mu = \frac{i \sum_\lambda g_{\mu\lambda} b_\lambda d_\mu}{i(\omega - \omega_\lambda) + \gamma} \tag{3-89}$$

Recalling that the photon number $n_\lambda$ is given by $b_\lambda^* b_\lambda$ and that

$$\frac{dn_\lambda}{dt} = b_\lambda \frac{db_\lambda^*}{dt} + b_\lambda^* \frac{db_\lambda}{dt} \tag{3-90}$$

one can show, from (3-86a), that

$$\frac{dn_\lambda}{dt} = -2\kappa_\lambda n_\lambda + i \sum_\mu g_{\mu\lambda} b_\lambda \alpha_\mu^* - i \sum_\mu g_{\mu\lambda}^* b_\lambda^* \alpha_\mu \tag{3-91}$$

Substituting (3-89) in (3-91), one finds that

$$\frac{dn_\lambda}{dt} = -2\kappa_\lambda n_\lambda + \sum_{\lambda'} \sum_\mu \frac{g_{\mu\lambda} b_\lambda g_{\mu\lambda'}^* b_{\lambda'}^* d_\mu}{-i(\omega - \omega_\lambda) + \gamma} + \sum_{\lambda'} \sum_\mu \frac{g_{\mu\lambda}^* b_\lambda^* g_{\mu\lambda'} b_{\lambda'} d_\mu}{i(\omega - \omega_\lambda) + \gamma} \tag{3-92}$$

That there are terms with both $\lambda$'s and $\lambda'$'s indicates that there is beating between the various modes. This beating is one of the things that is ignored in rate equation treatments, and therefore for the present we assume $\lambda = \lambda'$ to obtain

$$\frac{dn_\lambda}{dt} = -2\kappa_\lambda n_\lambda + \sum_\mu W_{\mu\lambda} d_\mu n_\lambda \tag{3-93}$$

where the $W_{\mu\lambda}$ is given by

$$W_{\mu\lambda} = \frac{2\gamma|g_{\mu\lambda}|^2}{(\omega - \omega_\lambda)^2 + \gamma^2} \tag{3-94}$$

A similar treatment of the equation of $d_\mu$ leads to

$$\frac{dD}{dt} = \frac{D_0 - D}{\tau_d} - 2 \sum_\lambda \sum_\mu W_{\mu\lambda} d_\mu n_\lambda \tag{3-95}$$

where the $D$ and $D_0$ are defined by

$$D = \sum_\mu d_\mu \tag{a}$$

$$D_0 = \sum_\mu d_0 \tag{b} \quad (3\text{-}96)$$

Often one denotes

$$\Gamma_\lambda = \sum_\mu W_{\mu\lambda} d_\mu \tag{3-97}$$

allowing one to write

$$\frac{dn_\lambda}{dt} = (\Gamma_\lambda - 2\kappa_\lambda)n_\lambda \tag{a}$$

$$\frac{dD}{dt} = \frac{D_0 - D}{\tau_d} - 2 \sum_\lambda \Gamma_\lambda n_\lambda \tag{b} \quad (3\text{-}98)$$

Let us consider the rate equations for a single mode laser in greater detail. For a single mode laser, one can ignore the subscript $\lambda$. Let us further assume that the coupling constant is independent of $\mu$ (homogeneously broadened medium) and therefore can be simply expressed as $g$. In this case, the rate equations can be written as

$$\frac{dn}{dt} = Gn \qquad \text{(a)}$$

$$\frac{dD}{dt} = \frac{D_0 - D}{\tau_d} - 2\Gamma_n \qquad \text{(b)} \quad \text{(3-99)}$$

where $G = \Gamma - 2\kappa$ is the net optical gain of the medium. Now the $\Gamma$ will exhibit frequency dependence. Indeed, the $\Gamma$ is expressible as

$$\Gamma = \frac{2\gamma g^2 (N_b - N_a)}{(\omega - \omega_\lambda)^2 + \gamma^2} \qquad \text{(3-100)}$$

where $N_b$ is the number of atoms in the upper state and $N_a$ is the number of atoms in the lower state and the subscript $\lambda$ is needed here to distinguish the operating frequency $\omega_\lambda$ from the line center frequency $\omega$. If one plots $\Gamma$ as a function of $\omega_\lambda$, one finds the familiar Lorentzian lineshape pictured in Figure 3.14. The idea is that, for an inverted medium in which $N_b > N_a$, the gain will be strongest at the line center $\omega_\lambda = \omega$. Unfortunately, this gain maximum may not coincide with one of the Fabry-Perot modes and lasing may have to occur away from the line center. The important quantity for the photon number evolution, however, is the net gain function $G = \Gamma - 2\kappa$. For example, we could consider a case in which $0 < \Gamma < 2\kappa$, and at $t = 0$ there are $n_0$ photons in the cavity and the inversion value is $D_i$. The evolution is sketched in Figure 3.15. The point is that nothing too exciting happens. The case where $G$ is greater than zero is really the one of interest, and indeed it is one that we shall return to in a later paragraph.

We now wish to turn to the question of the stability of the solutions of the semiclassical system as a function of the pump parameter $D_0$. We saw above that, if the net gain in the rate equations were negative, the only stable solution for the field would be zero. We now wish to consider the stability of the whole system. We again wish to consider a single mode of a homogeneously broadened medium whose $D$ value has stabilized to the quasi-

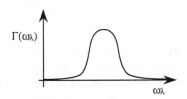

**FIGURE 3.14**    Variation of $\Gamma$ with $\omega_\lambda$.

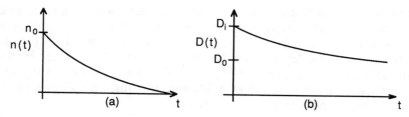

**FIGURE 3.15** Evolution of (a) the photon number $n(t)$ and (b) inversion parameter $D(t)$ for a case where $0 < \Gamma < 2\kappa$.

equilibrium pump value $D_0$. The system of (3-86) thereby reduces to

$$\frac{db}{dt} = (-i\omega_\lambda - \kappa)b - igS \qquad \text{(a)}$$

$$\frac{dS}{dt} = (-i\omega_\mu - \gamma)S + iD_0gb \qquad \text{(b)} \quad \text{(3-101)}$$

where the $S$ is defined as

$$S = \sum_\mu \alpha_\mu \qquad \text{(3-102)}$$

Clearly, a possible solution for the system is $b = S = 0$ since it is a homogeneous linear system. To perform a stability analysis, what one generally does is to check for the boundedness of small perturbations about an operating point. Because the perturbations are initially small, the system can be linearized in the perturbation parameters. For our problem, this corresponds to assuming a solution

$$b = b_0 + \delta b \qquad \text{(a)}$$

$$S = S_0 + \delta S \qquad \text{(b)} \quad \text{(3-103)}$$

where, in our case, $b_0 = S_0 = 0$ and the $\delta b$ and $\delta S$ are to be considered infinitesimally small. The equations satisfied by the perturbations are therefore found to be

$$\frac{d\delta b}{dt} = (-i\omega - \kappa)\delta b - ig\delta S \qquad \text{(a)}$$

$$\frac{d\delta S}{dt} = (-i\omega - \gamma)\delta S + igD_0\delta b \qquad \text{(b)} \quad \text{(3-104)}$$

These equations look strangely like the ones we started with because, indeed, they are. This arose from the fact that the original system was linear. Had we also expanded the $D$ equation for $D(t) = D_0 + \delta D$, we would have found a lot of simplification. Perhaps the interested reader would like to perform this transformation to convince himself that stability analysis works. At any rate, the next step is to assume an exponential solution (which is a solution of any constant coefficient linear system) of the form

$$\delta b = \delta b_0 e^{(-i\Omega + \xi)t} \tag{a}$$

$$\delta S = \delta S e^{(-i\Omega + \xi)t} \tag{b} \quad (3\text{-}105)$$

and find under what conditions the $\xi$ are negative and therefore the perturbations are damped. Plugging the ansatz of (3-105) into (3-104), one can obtain the matrix relation

$$\begin{bmatrix} -i\Omega + \xi + i\omega + \kappa & ig \\ -igD_0 & -i\Omega + \xi + i\omega + \gamma \end{bmatrix} \begin{pmatrix} \delta b_0 \\ \delta S_0 \end{pmatrix} = 0 \tag{3-106}$$

The $\xi$ are determined from the condition that the determinant of the system must be zero (as it is a homogeneous system). The roots of this relation become

$$\xi_{\pm} = -\frac{\kappa + \gamma}{2} \pm \sqrt{\frac{(\kappa - \gamma)^2}{4} + g^2 D_0} \tag{3-107}$$

with the corresponding condition for stability that

$$D_0 < \frac{\kappa \gamma}{g^2} \tag{3-108}$$

Let us compare the condition (3-108) with what we found in the rate equation case, that is that interesting things happened when $\Gamma > \kappa$. If one uses (3-100) with the assumption that one is operating at the line center $\omega$, one will immediately find that $\Gamma > \kappa$ is identical with (3-108). Now one of the problems with stability analysis is that it will tell us when things are unstable, but not how they behave when they go unstable. The rate equations will do so, however, in an approximate way at least. Equations (3-100) are terribly nonlinear despite their linear appearance due to the fact that the $\Gamma$ of equation (3-100) is linear in $D$. However, the expanding and linearizing about the operating point trick that we just used above could work for us again here. For example, one can find steady state values for $n$ and $D$ by setting the left-hand sides of (3-99) to zero to obtain

$$n_{SS} = \frac{D_0 - D_{SS}}{2\Gamma_{SS}\tau_d} \qquad \text{(a)}$$

$$D_{SS} = \frac{\kappa\gamma}{g^2} \qquad \text{(b)} \quad \text{(3-109)}$$

where the subscripts SS on $\Gamma_{SS}$ indicate that $\Gamma$ must be evaluated at $D_{SS}$. One could then expand the system about this operating point, as was done in Equation (3-103), and linearize. One would then obtain a linear system which could be solved as an initial value problem. The system might be written in the form

$$\frac{d\delta n}{dt} = Wn_{SS}\delta D + G\delta n \qquad \text{(a)}$$

$$\frac{d\delta D}{dt} = -\frac{\delta D}{\tau_d} - 2\delta\Gamma n_{SS} - 2\Gamma_{SS}\delta n \qquad \text{(b)} \quad \text{(3-110)}$$

For a problem of gain switching at time zero, one might want to take an initial condition such as $\delta n = -n_{SS}$ and $\delta D = 0$. Clearly, the initial negative value of $\delta n$ will act as a pump of $\delta D$ in (3-110b), which in turn will cause $Wn_{SS}\delta D$ to pump (3-110a). When $\delta n$ becomes positive, the effect will reverse. Relaxation oscillations are the result. Sketches of this behavior are contained in Figure 3.16. The details of the analysis are left to the reader.

### 3.3 Subthreshold Behavior

One rather strange thing which has yet to be commented upon is the fact that, $G \leq 0$, there was no radiation whatsoever. Yet we know well that light bulbs

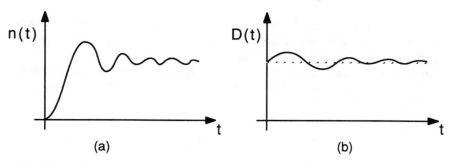

(a)                                           (b)

**FIGURE 3.16**    Sketches of (a) $n(t)$ and (b) $D(t)$ for the case of relaxation oscillation in a gain switched laser.

work, even though they are thermal in natural and can therefore never reach inversion or threshold. The point here really is that we have everywhere ignored spontaneous emission. For a light bulb, this is the primary source of the electric field. Although we do not want to derive in detail how to fix this problem, especially seeing that this would require field quantization and therefore the theory of quantum electrodynamics, we do wish to know the solution to it, which is embodied in the following modified semiclassical system

$$\frac{db_\lambda}{dt} = (-i\omega_\lambda - \kappa_\lambda)b_\lambda - i\sum g^*_{\mu\lambda}\alpha_\mu + F_\lambda(t) \tag{a}$$

$$\frac{d\alpha_\mu}{dt} = (-i\omega - \gamma)\alpha_\mu + i\sum g_{\mu\lambda}b_\lambda d_\mu + \Gamma_{\mu-}(t) \tag{b}$$

$$\frac{dd_\mu}{dt} = \frac{d_0 - d_\mu}{\tau_d} + 2i\sum(g^*_{\mu\lambda}\alpha_\lambda b^*_\lambda - g_{\mu\lambda}\alpha^*_\lambda b_\lambda) + \Gamma_d(t) \tag{c} \quad (3\text{-}111)$$

where the three new terms are zero-mean Langevin terms, meaning that they are random noise generators which take into account the spontaneous emission process which we have everywhere ignored in our semiclassical analysis. Quite generally, one ignores the $\Gamma_{\mu t}(t)$ and $\Gamma_d(t)$ for the reason that, in a laser at threshold, the number of carriers can be $10^8$, while the number of photons may be less than $10^4$. Therefore the $F_\lambda(t)$ will have significantly larger effect on the photon dynamics than the others would have on the carrier dynamics. Indeed, we shall ignore those two functions in all that follows. To fully define $F_\lambda$, one must define its moments. The second moments of it are given by

$$\langle F^*_\lambda(t)F_{\lambda'}(t')\rangle = 2\kappa_\lambda n_\lambda \delta(t - t')\delta_{\lambda\lambda'} \tag{a}$$

$$\langle F_\lambda(t)F^*_{\lambda'}(t')\rangle = 2\kappa_\lambda(n_\lambda + 1)\delta(t - t')\delta_{\lambda\lambda'} \tag{b} \quad (3\text{-}112)$$

To make the meaning of the noise terms clear, we wish to embark on one more detailed calculation before closing this chapter. The point here is to derive an equation that is valid near threshold and try to see how the Langevin terms modify the behavior of the field. Right from the beginning, we shall assume single mode operation and homogeneous broadening to obtain the Maxwell-Bloch system in the form

$$\frac{db}{dt} = (-i\omega_\lambda - \kappa)b - ig^*\alpha_\mu + F(t) \tag{a}$$

$$\frac{d\alpha_\mu}{dt} = (-i\omega - \gamma)\alpha_\mu + igbd_\mu \tag{b}$$

$$\frac{dd_\mu}{dt} = \left(\frac{d_0 - d_\mu}{\tau_d}\right) + 2i\,(g^*b^*\alpha_\mu - gb\alpha_\mu^*) \qquad \text{(c)} \quad \text{(3-113)}$$

where the noise generator is only included in the (a) equation for reasons discussed above. We begin by assuming that $\gamma$ is so large that it swamps the derivative term as well as the $\omega$ term in order to obtain

$$\alpha_\mu^{(1)} = \frac{igbd_0}{\gamma} \qquad \text{(3-114)}$$

where it further has been assumed that $d$ has come to steady state. We now want to assume that $d_\mu$ is so slowly varying that one can safely ignore its derivative. Doing this and substituting (3-114) in (3-113c), one obtains

$$d_\mu = d_0 \left(1 - \frac{4\tau_d|g|^2|b|^2}{\gamma}\right) \qquad \text{(3-115)}$$

which indicates that $d_\mu$ linearly decreases with increasing light intensity. One can use (3-115) to improve (3-114) by writing

$$\alpha_\mu^{(2)} = \frac{igbd_0}{\gamma} \left(1 - \frac{4\tau_d|g|^2|b|^2}{\gamma}\right) \qquad \text{(3-116)}$$

Using (3-116) back in (3-113a), one finds the equation we wanted — that is,

$$\frac{db}{dt} = (-i\omega - \kappa)b + \frac{g^2bD_0}{\gamma}\left(1 - \frac{4\tau_d|g|^2|b|^2}{\gamma}\right) + F(t) \qquad \text{(3-117)}$$

which is almost of the form of Van der Pol's equation. To get it to this form, we take

$$b(t) = B(t)e^{-i\omega t} \qquad \text{(3-118)}$$

to obtain

$$\frac{dB}{dt} = -\kappa B + \frac{g^2 D_0}{\gamma} B \left(1 - \frac{4\tau_d|g|^2|B|^2}{\gamma}\right)$$

$$+ F(t)e^{i\omega t} = \frac{G}{2} B - C|B|^2 B + F(t)e^{i\omega t} \qquad \text{(a)}$$

$$\langle F^*(t)F(t')\rangle \cong 2k[n_{\text{th}} + n_{\text{sp}}]\delta(t - t') \qquad \text{(b)} \quad \text{(3-119)}$$

where $G$ is just the net gain of (3-99) and one can consider $C$ as being defined by (3-119). The $n_{th}$ and $n_{sp}$ of (3-119b) are the number of photons due to the thermal and spontaneous generation mechanisms, respectively.

To bring out the meaning of (3-119), we wish to take a short excursion through the wonderful world of marbles. A marble of mass $m$ which moves in a bowl whose surface is defined by a function $V(x)$ has an energy given by

$$E_{total} = E_{kinetic} + E_{potential} = \frac{1}{2} m\dot{x}^2 + V(x) \qquad (3\text{-}120)$$

where it has tacitly been assumed that the potential walls (surfaces of the bowl) are not steep. Assuming the system is lossless, this $E_{total}$ should be a constant. Taking a derivative of the expression of (3-120) gives the relation

$$m\ddot{x} = -\frac{\partial V(x)}{\partial x} \qquad (3\text{-}121)$$

for the equation of motion of the marble. Damping can easily be included in this model by writing

$$m\ddot{x} + \gamma\dot{x} = -\frac{\partial V(x)}{\partial x} \qquad (3\text{-}122)$$

Depending on the relative size of $\gamma$, different forms of behavior can be observed. Some of them are sketched in Figure 3.17. If we limit consideration to overcritically damped motion, we could express our equation of motion in the form

$$\dot{x} = -\frac{\partial V(x)}{\partial x} \qquad (3\text{-}123)$$

where the damping coefficient has been absorbed into the $V(x)$. Now let us

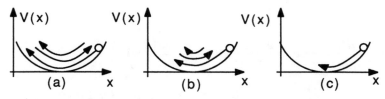

**FIGURE 3.17**    Motion of a marble in a bowl $V(x)$ with (a) no damping, (b) small damping, (c) overcritical damping.

consider a $V(x)$ of the form

$$V(x) = \frac{-G(D_0)}{4} x^2 + \frac{C(D_0)}{4} x^4 \tag{3-124}$$

where the functions $G(D_0)$ and $C(D_0)$ are given by

$$G(D_0) = 2 \frac{g^2}{\gamma} D_0 - 2\kappa \tag{a}$$

$$C(D_0) = 4 \frac{g^2}{\gamma^2} \tau_d |g|^2 D_0 \tag{b} \tag{3-125}$$

and their behaviors are sketched in Figure 3.18. Now the shape of the potential is also going to be affected by the value of $D_0$. This situation is graphically depicted in Figure 3.19. For negative values of $D_0$, the potential is essentially a harmonic oscillator potential close to the origin, but it eventually falls off due to the $x^4$ values for large values of $x$. When $D_0$ exceeds zero but is less than the threshold value, which we earlier denoted as $D_{ss}$, the potential still appears like that of a harmonic oscillator close enough to the origin. When $D$ takes on its threshold value, the coefficient of $x^2$ becomes zero and the potential takes on only the $x^4$ behavior. Any value $D_0$ exceeding threshold, however, will cause two broad minima to appear symmetrically about the origin. In this case, the marble will no longer come to rest at the origin but at one of these minima.

The problem we now face is to apply the above picture to our spontaneous emission driven field equation of (3-111). As this equation is complex, the coordinate $B$ is actually 2-dimensional, and therefore our marble will move in a 2-dimensional potential. One can make the substitution

$$B(t) = r(t) e^{i\phi(t)} \tag{3-126}$$

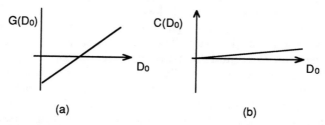

(a)                          (b)

**FIGURE 3.18**    Plots of the behaviors of $G(D_0)$ and $C(D_0)$ as functions of $D_0$.

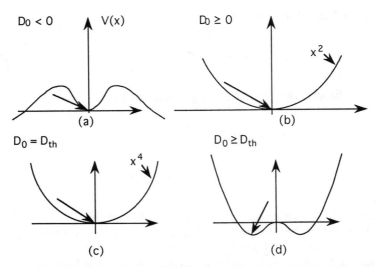

**FIGURE 3.19**   Sketches of the potential $V(t)$ for various values of the parameter $D_0$, from $D_0 = 0$ up to above the $D_{ss}$ value that was discussed in connection with the rate equations.

in the field equation to obtain the two equations

$$\frac{dr}{dt} = \frac{G}{2}r - Cr^3 + F_r(t) \qquad \text{(a)}$$

$$\frac{d\phi}{dt} = F_\phi(t) \qquad \text{(b)} \quad \text{(3-127)}$$

where the definitions of $F_r(t)$ and $F_\phi(t)$ should be self-explanatory. The idea here is that the potential has azimuthal symmetry, as there is no restoring force in the $\phi$ direction. The pictures of Figure 3.19, therefore, still directly apply if one just imagines these shapes as having azimuthal symmetry. The motion of the marble will be a little more complicated here due to the extra coordinate and the noise term. Well below threshold, the marble will have an average coordinate of $r = 0$, although the noise source will cause this coordinate to have a standard deviation. The noise source acts like a very young soccer player running around this azimuthally symmetric valley kicking the marble at random periods in totally random directions but always with exactly the same strength. As the grass in the valley is very deep, the marble's motion is damped in the radial direction. However, the grass must have been cut in layers, as the motion around the valley walls is undamped (see (3-16)). As the system is taken over threshold, however, the shape of the valley

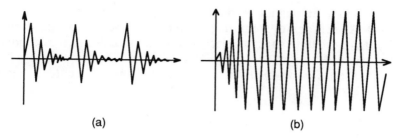

(a)                              (b)

**FIGURE 3.20**    Sketch of the evolution of spontaneous emission fluctuations (a) below threshold, and (b) above threshold.

changes radically, as the minimum moves out and forms a ring about the center. No matter where the marble was when threshold occurred, the little soccer player running around after it will give it enough motion that it will "sniff out" the new minimum quite rapidly. This is the phenomenon of laser turn-on. The situations below and above threshold can be visualized in terms of the sketches of Figure 3.20. Below threshold, a spontaneous emission event (a kick of the ball up the hill) is reasonably rapidly damped out (the ball rolls down the embankment). As one nears threshold, the time taken to damp out increases (the marble takes longer to fall down the less steep slope). Above threshold, however, a single spontaneous event will be amplified up to a steady state value (a single kick will cause the ball to roll down the embankment to some position in the minimum ring). The thermal light source corresponds to a system below threshold. The light intensity is due only to the fluctuations of the field value around a zero mean. This is clearly an inefficient way to generate light when compared to an above threshold biased laser where the field has small fluctuations about a large mean value. The lack of damping of the $\phi$ equation, however, dictates that even the laser field cannot be purely monochromatic, as the time varying phase (the rolling of the marble around the hill) will cause frequency fluctuations. In the next chapter, we shall see that this situation is even more serious in a semiconductor laser.

## Reference

H. Haken, 1985. *Light,* Volume 2, *Laser Light Dynamics.* North-Holland Publishing Company, Amsterdam.

## PROBLEMS

1. A handy technique for finding the binding energy and Bohr radius of hydrogen was introduced in Equations (3-22)–(3-24). Apply this

method to finding the binding energy and Bohr radius of:

(a) A positronium atom, that is, an electron bound to a positron

(b) An exciton, that is, an electron bound to a hole in a semiconductor of index $n$

(c) A system in which the masses $m_1$ and $m_2$ feel a potential energy given by

$$E_{\text{pot}} = -g\,\frac{e^{-\mu r}}{r}$$

(d) A system in which the masses $m_1$ and $m_2$ feel a potential $E_{\text{pot}} = -g\,\dfrac{1}{r^3}$ (a 2-quark system?)

2. Consider a particle moving in an attractive potential of $-g^2/r^N$, with a kinetic energy of $p^2/2m$. Find a Bohr radius and a binding energy for this particle. What is wrong with this problem when $N \geq 2$?

3. Consider a 1-dimensional Shroedinger equation, $H_0\psi = E\psi$, with a Hamiltonian $H_0$, such that

$$\frac{d^2\psi}{dx^2} + \frac{2M}{\hbar^2}\,(E - V)\,\psi = 0$$

in the infinite square well potential, as depicted in Figure 3.21.

(a) Find the modes $\psi_n$ of this potential. The label $n$ of these modes indicates the number of nodes of the wavefunction. The modes should be normalized such that

$$\int_{-\infty}^{\infty} \psi_m(x)\psi_n(x)\,dx = \delta_{nm}$$

where $\delta_{nm}$ is the Kronecker delta function.

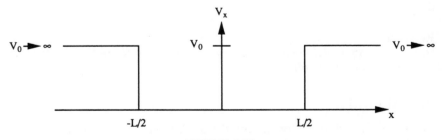

**FIGURE 3.21**

(b) Find $x_{nm} = \langle x \rangle$ between the above derived modes, where

$$x_{nm} = \int \psi_n(x) x \psi_m(x) \, dx$$

Note: This value gives you the effective dimensionless dipole moment between the modes $n$, $m$.

4. The coupled equations for the level occupations of a 2-level system derived in text ((3-44)) were

$$\frac{\partial C_a}{\partial t} = \frac{\mathscr{P} \cdot \mathbf{E}(t)}{i\hbar} C_b e^{-i\omega t} \qquad (a)$$

$$\frac{\partial C_b}{\partial t} = \frac{\mathscr{P} \cdot \mathbf{E}(t)}{i\hbar} C_a e^{-i\omega t} \qquad (b)$$

(a) Find the "rotating wave" wave approximation to these equations. That is, assume

$$\mathbf{E}(t) = \mathbf{E}_0 \, 2 \cos(\omega_\lambda t) = \mathbf{E}_0(e^{-i\omega_\lambda t} + e^{i\omega_\lambda t})$$

and eliminate rapidly varying exponentials.
(b) Solve the equations obtained in (a) by assuming $C_a(t) = e^{i\mu t}$ and solving for $\mu$.
(c) Find the resulting $\mathbf{p}^{(+)}$ for the $C_a$ and $C_b$ of (b)
(d) Obtain $\mathbf{p}^{(+)}$ as a power series in $\mathscr{P} \cdot \mathbf{E}_0/\hbar\omega$.
5. Say that we consider a system to consist of only two levels, that is, that any state of the system $\psi(x,t)$ is expressible as

$$\psi(x,t) = C_0(t)\psi_0(x) + C_1(t)\psi_2(x)$$

Also, consider that the system to be acted on by a perturbation, such that the total system Hamiltonian becomes

$$H(x,t) = H_0(x) = xe \, E_0 \cos \omega t$$

(a) Find the coupled equations satisfied by $C_0$ and $C_1$.
(b) Solve for $C_0(t)$ and $C_1(t)$ subject to $C_0(0) = 1$, $C_1(0) = 0$, and in the approximation that $\hbar\omega \sim \Delta E \approx E_1 - E_0$. Assume higher harmonics than the first of $\omega$ are negligible.
(c) Find and plot the probability that the system is in state 1 and state 2.
(d) What does the probability $P(x,t)$ look like for several values of the time?

6. Say we have a 1-dimensional, 2-level system with modes

$$\psi_c = \frac{1}{\sqrt{L}} \cos kx$$

$$\psi_s = \frac{1}{\sqrt{L}} \sin kx$$

where $-L < x < L$ and $L \to \infty$. We wish to calculate the intrinsic dipole moment

$$\mathcal{P} = \int_{-L}^{L} \psi_c \, x \psi_s \, dx$$

for this system. Make this calculation and comment on any mathematical problems associated with the calculation. Recall that

$$\int x \sin x \, dx = \sin x - x \cos x$$

How would one fix up the mathematical problems in a physically reasonable way?

7. It can be shown that a component of the scalar microscopic polarizability, $p_i$, satisfied the almost classical equation

$$\ddot{p}_i + \omega^2 p_i = KE_i d(t)$$

where $K$ is some constant. In the classical, linear limit, the positive frequency part of this $p_i$ can be related to $\epsilon(\omega)$ of this material by

$$(\epsilon(\omega) - \epsilon_0) = \frac{N p_i^{(+)}}{E_i}$$

where $N$ is a density of atoms. Assuming $E_i$ and $d(t)$ are not affected (to first order) by changes in $\epsilon(\omega)$, find the spectrum of $\epsilon(\omega)$ for $E_i = \psi_i e^{-i\omega_i t}$ and for:
(a) $d(t) = \text{constant} = d_0$
(b) $d(t) = d_0 + d_1 \, \text{rect} \, (t/\tau)$
(c) $d(t) = d_0 + d_1 \sin \omega_d t$
(d) In each of the above three cases describe what effect the calculated $\epsilon(\omega)$ would have on the spectrum of a wave propagating through that material.

**8.** In many books, attention is given to 2-level systems. In general, there will be $N$ levels.

   **(a)** Find the dynamical equations for the coefficients $C_i(t)$ of the expansion

$$\psi(\mathbf{r},t) = \sum_{i=0}^{N} C_i(t)u_i(\mathbf{r})e^{-i\frac{E_i}{k}t}$$

   where the time variation is due to the perturbation of an applied electric field $E_0 e^{-i\omega t}$.

   **(b)** Use slowly varying and weakly perturbed approximations to solve the dynamical system for the $C_i$. You can assume that $C_0(t) \sim C_0(0)$, i.e. the system stays essentially in its ground state.

   **(c)** Find an expression for the microscopic dipole moment $\mathbf{p}_\omega$.

   **(d)** Find an expression for the index of refraction $n^2(\omega)$. Ignore local field corrections.

**9.** Recall the following system of equations:

$$\frac{\partial b}{\partial t} = (-i\omega_\lambda - \kappa)b - ig\alpha$$

$$\frac{\partial \alpha}{\partial t} = (-i\omega_\alpha - \gamma)\alpha + igbd$$

$$\frac{\partial d}{\partial t} = \frac{d_0 - d}{\tau} + 2i(gb^*\alpha - gb\alpha^*).$$

A standard trick used to find a solution to these equations is to assume the damping $\gamma$ is large, perturbatively solve the $\alpha$ equation, substitute this solution into the (assumed slowly varying) $d$ equation, substitute this result back into the $\alpha$ equation and use this result to obtain a differential equation in $b$ alone. Here, assume $\kappa$ is the large parameter, and carry out a similar program to obtain an equation for $\alpha$ alone. What does the resulting differential equation tell you?

**10.** Recall the semiclassical laser equations for a single-mode laser:

$$\dot{b} = (-i\omega - \kappa)b - i\sum_\mu g_{\mu\lambda}^* \alpha_\mu$$

$$\dot{\alpha}_\mu = (-\overline{\omega}_\mu - \gamma)\alpha_\mu + ig_{\mu\lambda}d_{\mu b}$$

$$\dot{d}_\mu = \frac{d_0 - d_\mu}{T} + 2i(g_{\mu\lambda}^* \alpha_\mu b^* - g_{\mu\lambda}\alpha_\mu^* b)$$

Assuming a solution of the form

$$b = Be^{-i\Omega t}$$

$$\alpha_\mu = A_\mu e^{-i\Omega t}$$

$$d_\mu = \text{constant}$$

show that

$$\Omega = \frac{\overline{\omega}\kappa + \omega\gamma}{\lambda + \kappa}.$$

11. Assume a solution of the form

$$b = Be^{-i\Omega t}$$

$$d_\mu = \text{constant}$$

to the single-mode laser equations. From this initial assumption, calculate an $\alpha_\mu^{(1)}$; use this to calculate an updated $d_\mu^{(1)}$ and use this to again update $\alpha_\mu^{(2)}$. From the $\alpha_\mu^{(2)}$, find the nonlinear equation satisfied by $b$. Note that this equation has the form of a Van der Pol oscillator equation in the slowly varying limit.

12. A Van der Pol-type equation for the electric field is derived in the text. The highest order term in this equation was third order, or $|b_\lambda|^2 b_\lambda$.
    (a) Go back to the derivation and find the fifth-order term.
    (b) As was done in the test, find the potential which corresponds to the equation with the fifth-order correction.
    (c) Sketch the potential of (b).
    (d) Are there higher-order phase transitions associated with the sixth-order potential? What would such a transition mean?

13. In deriving the rate equations, one generally adiabatically eliminates the polarizability from the semiclassical system

$$\dot{b}_\lambda = (-i\omega_\lambda - \kappa_\lambda)b_\lambda - i \sum g_{\mu\lambda}^* \alpha_\mu$$

$$\dot{\alpha}_\mu = (-i\omega - \gamma)\alpha_\mu + i \sum_\lambda g_{\mu\lambda}b_\lambda d_\mu$$

$$\dot{d}_\mu = \frac{d_0 - d_\mu}{t_d} + 2i \sum_\lambda g_{\mu\lambda}(b_\lambda^* \alpha_\mu - b_\lambda \alpha_\mu^*)$$

by formally solving the $\alpha_\mu$ equation in the form

$$\alpha_\mu(t) = e^{-i\omega t}e^{-\gamma t}i\sum_\lambda g_{\mu\lambda}\int_0^t dt'\,B_\lambda(t')d_\mu(t')e^{i(\omega-\omega_\lambda)t'}e^{+\gamma t}$$

and then assuming $\gamma$ is large enough that $B_\lambda$ and $d_\mu$ can be removed from the integral and evaluated at $t$.

(a) By Taylor expanding $B_\lambda$ and $d_\mu$ for $t'$ near $t$, find the next terms in the adiabatic expansion for $\alpha$.

(b) Find the rate equations with these correction terms.

(c) What effects will these forms have on the solutions to the rate equations? In particular, how will they effect relaxation oscillations.

14. A classical dispersion model was used in the text to derive the susceptibility $\hat{x}$. This function was plotted in the text on axes $x_r$ and $x_i$ as a function of $\omega$, the frequency. What is the equation of the resulting shape in the $x_r$, $x_i$ plane?

15. The rate equation

$$\frac{d\Delta N}{dt} = -W_{12}\Delta N - \frac{(\Delta N - \Delta N_{eq})}{\tau_{NR}}$$

was derived in the text. Find the complete solution to this differential equation for $W_{12}$ and $\tau_{NR}$ constant.

16. Consider a rate equation as in problem 15 above. Say that one can generate a $W_{12}$ such that

$$W_{12} = W_a + W_b\cos\omega t$$

where $W_b \ll W_a$. For such pumping, one would expect to find that

$$\Delta N = N_a + N_b\cos(\omega t + \delta)$$

to first order. Find $N_a$, $N_b$ and $\delta$. Can this technique be used as a material diagnostic? Explain.

17. Recall the rate equations

$$\frac{dn}{dt} = DWn - 2\kappa n$$

$$\frac{dD}{dt} = \frac{1}{T}(D_0 - D) - 2WDn$$

where $n$ is the photon number and $D$ the inversion.

(a) Derive the threshold condition by assuming adiabatic changes in the pumping and that $2TWn \ll 1$ for usual laser pumping.

(b) Assuming the steady state for $D$ found in a.) above, derive an equation for $n$ of the form

$$\frac{dn}{dt} = an - bn^2$$

and find $a$ and $b$.

(c) Show the above equation for $\dfrac{dn}{dt}$ can be solved by

$$n(t) = \frac{ac\ e^{a(t-t_0)}}{1 + bc\ e^{a(t-t_0)}} \qquad \text{for } a > 0$$

$$c = \frac{n(t_0)}{a - b\ n(t_0)}$$

and

$$n(t) = \frac{|a|\ c\ e^{a(t-t_0)}}{1 - bc\ e^{a(t-t_0)}} \qquad \text{for } a < 0$$

18. In Chapter 3, some attention is given to relaxation oscillations. In fact, Equation (3-110) and the accompanying discussion sets up the relaxation oscillation problem as an initial value problem. Solve these equations and sketch the resulting behavior for a few "reasonable" sets of parameters.

19. Consider the system of rate equations for an $N$ level system

$$\frac{dN_n}{dt} = -\sum_m W_{nm}\ (N_n - N_m)$$

$$+ \sum_m -w_{nm}\ N_n + w_{mn}\ N_m$$

where $W_{nm}$ represents the stimulated (field driven) transition rate, $w_{nm}$ represents the spontaneous (thermally driven) transition rate and $N_n$ represents the number of atoms in the $n^{\text{th}}$ excited state.

(a) What conditions must be satisfied by the $W$'s and $w$'s if the number of atoms in the system is to be conserved.

(b) If we are to demand that, in thermal equilibrium, with no applied field, the $N_n$'s must satisfy

$$N_n = \frac{N_{\text{total}}\, e^{-\epsilon_n/k_B T}}{\Sigma i\, e^{-\epsilon_i/k_B T}} = N_{n0}$$

where $N_{\text{total}}$ is the total number of atoms in the system, $\epsilon_i$ is the $i^{\text{th}}$ energy level, $k_B$ is Boltzman's constant and $T$ the temperature, what are the $w_{nm}$'s?

(c) In the low frequency approximation, one takes $N_n \sim N_m \sim N_{n0} \sim N_{m0}$ for all $m, n$. Use this approximation to find the approximate form of the rate equation

$$2\frac{d}{dt}N_n(t) = -2\sum_m \left[2W_{nm}\,\Delta N_{nm}(t) - \frac{\Delta N_{nm}(t) - \Delta N_{0,\,nm}}{2\tau_{l,\,nm}}\right]$$

where

$$\Delta N_{nm} = N_n - N_m,\ \Delta N_{0,\,nm} = N_{\text{total}}\frac{e^{-\epsilon_n/k_B T} - e^{-\epsilon_m/k_B T}}{\sum_i e^{-\epsilon_i/k_B T}},$$

and $\tau_{l,\,nm}$ is the decay time for the transition $n \to m$.

(d) Write out the above equations for a three-level system.

(e) Consider that in the system of (d) one pumps the $1 \to 3$ transition and ignores the upward rates $w_{12}, w_{13}$ with respect to the downward rates (high frequency approximation). What additional condition is necessary to achieve an inversion of the 2,1 population?

20. Consider the following system of equations which represent the time development of the inversion and photon number in a 2-level system:

$$\frac{dD}{dt} = \frac{D_0 - D}{T} - 2WDn$$

$$\frac{dn}{dt} = DWn - 2\kappa n$$

Say that at some time $t = 0$, the term $\dfrac{D_0 - D}{T}$ is instantaneously increased to some value $\dfrac{\Delta D}{T}$, and (by some miracle) maintained at this value. Assume $D_0$ $(t < 0)$ is truly of thermal origin (i.e. $D_0 < 0$), but $\Delta D > 0$.

(a) Find the "pre-threshold" behavior of $D(t)$.
(b) Find the time $t_0$ at which inversion is achieved.
(c) Find the approximate post-threshold behaviors of $D(t)$, $n(t)$.
(d) Sketch $D(t)$ and $n(t)$ for $t > 0$.

21. Consider the gross approximation to the laser rate equations

$$\frac{dI}{dt} = (g^2 D - 2K) I$$

$$\frac{dD}{dt} = \frac{D_0 - D}{\tau_d}$$

subject to the initial conditions that

$$I(0) = 1$$
$$D(0) = D_i$$

Decouple this system to obtain an equation for $I$ alone. Is there any limit in which you can solve this equation? If you can, solve it in this limit.

22. Langevin noise sources can be modeled as being of the form

$$f(t) = \sum_i \delta(t - t_i) e^{i\phi_i}$$

where the $t_i$ are Poisson random variables and the $\phi_i$ are uniformly distributed. The electric field equation is generally written in the form

$$\dot{b}_\lambda = (-i\omega_\lambda - \kappa_\lambda) b_\lambda - ig^* \alpha + f(t)$$

In what follows, assume that the spontaneous emission rate is less than $\kappa_\lambda$, one over the photon lifetime.

(a) Solve these equations assuming that $\alpha = -\dfrac{a}{ig^*} b$. The solution should be in the form of an infinite sum.
(b) Sketch what the solution $(\mathrm{Re}\,(b_\lambda))$ may look like taking $a = 0$.
(c) Sketch the solution for a couple values of $\alpha = i\alpha_i$ where $0 < a < K_\lambda$.
(d) What happens when $\alpha_i \to \kappa_\lambda/g$? How is a catastrophe averted when $\alpha_i$ attains this value?

23. Equations (3-127) express a system which should predict the dynamical behavior of a single mode laser not too far from threshold. An exact

solution to equation (3-127a) is given by

$$r(t) = \sqrt{\frac{Gh(t)}{1 + Ch(t)}}$$

with $h(t)$ given by

$$h(t) = \frac{r_0^2}{G - Cr_0^2} \exp[2G(t - t_0)]$$

where $r(t_0) = r_0$. Taking "reasonable" parameter sets, sketch some solutions for $r(t)$ which should correspond to cases depicted in Figure 3.19 and discussed in the paragraph following equation (3-127). Do the solutions agree? If not, why not?

# 4

# Semiconductor Lasers

In the last chapter, we considered a model for a two-level system in a resonator and derived the following set of dynamical equations for the system's time evolution:

$$\frac{db_\lambda}{dt} = (-i\omega_\lambda - \kappa_\lambda)b_\lambda - i \sum_\mu g_{\mu\lambda}^* \alpha_\mu + F_\lambda(t) \qquad \text{(a)}$$

$$\frac{d\alpha_\mu}{dt} = (-i\omega - \gamma)\alpha_\mu + i \sum_\lambda g_{\mu\lambda} b_\lambda d_\mu + \Gamma_\mu(t) \qquad \text{(b)}$$

$$\frac{dd_\mu}{dt} = \frac{d_0 - d_\mu}{\tau_d} + 2i \sum_\lambda (g_{\mu\lambda}^* b_\lambda^* \alpha_\mu - g_{\mu\lambda} b_\lambda \alpha_\mu^*) + \Gamma_d(t) \qquad \text{(c)} \quad \text{(4-1)}$$

To apply such a set of equations to any problem, however, it is necessary to know the parameters $\omega_\lambda$, $\kappa_\lambda$, $g_{\mu\lambda}$, $\omega$, $\gamma$, and $\tau_d$ with some accuracy as well as to understand the pump mechanism well enough to be able to calculate in some detail the dynamic behavior of $d_0$ with the external pump. An additional problem arises with semiconductors in that one cannot consider individual atoms in the material as the electrons are not localized in the lattice. Therefore, the index $\mu$ is no longer a valid index on which to sum. These issues will be taken up sequentially in the sections of this chapter. In the first section of the effective index method, calculation of $\omega_\lambda$, $\kappa_\lambda$, and the mode patterns $\mathbf{u}_\lambda(\mathbf{r})$ will be considered in some detail. The second section will deal with a very simple model of band structure for the purpose of describing how one can replace the atomic coordinate $\mu$ with the electron wave vector k. The third section will discuss the concepts of Fermi levels and quasi-Fermi levels with attention directed toward elucidating the meaning of the time constants $\tau_d$

and $\tau_\alpha = 1/\gamma$, as well as finding a dynamical model for the $d_0$ in terms of measurable parameters. The fourth section of the chapter will cast the Maxwell-Bloch equations in their normal semiconductor rate equation form while giving some discussion of the polarizability $\chi$ in the equations. The fifth and concluding section of the chapter will discuss something about laser structure and doping profile.

## 4.1 EFFECTIVE INDICES

In Chapter 3 (Equations (3-59)–(3-72)), it was pointed out that, if $\psi_n$ and $\beta_n(\omega)$ of a modal expansion of the form

$$\mathbf{E}(\mathbf{r},t) = \mathrm{Re}\left[\sum a_n \psi_n(x,y) e^{i\beta_n(\omega)Z} e^{-i\omega t}\right] \tag{4-2}$$

could be found for an arbitrary monochromatic field in an infinite wave-guide of transverse index distribution $n(x,y)$, then one could also find the $\omega_\lambda$ and $\mathbf{u}_\lambda(\mathbf{r})$ of the expansion

$$\begin{aligned}
\mathbf{E}(\mathbf{r},t) &= \mathrm{Re}\left[\sum_\lambda E_\lambda(t)\mathbf{u}_\lambda(\mathbf{r})\right] \\
&= \mathrm{Re}\left[\sum_\lambda a_\lambda(t)e^{-i\omega_\lambda t}\mathbf{u}_\lambda(\mathbf{r})\right]
\end{aligned} \tag{4-3}$$

of an arbitrary non-monochromatic field existing in a resonator filled with an identical dielectric waveguide. The process of generating the parameters of (4-3) from those of (4-2) is described in some detail in Chapter 3. What is not discussed is a method for finding the $\psi_n(x,y)$ and the $\beta_n(\omega)$. That will be a primary purpose of this section.

An archetypical problem of guidance is illustrated in Figure 4.1. The point here is that planar circuit technology is still our best bet for fabricating repeatable circuits of waveguides. The figure illustrates that the indiffused optical channel of our generic waveguide structure is rectangular and uniform in index. This is clearly unrealistic, but one needs to begin somewhere,

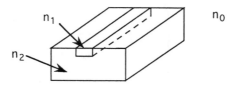

**FIGURE 4.1.** Typical waveguide guidance problem: guidance in an indiffused stripe of a substrate of index $n_1$.

so first we consider a "simple" problem, find out that it is essentially unsolvable, find a way to *somewhat* solve it, and then apply this "usable" solution method to try to say something about the structures in which we are interested.

In light of the above, let us turn our attention to the guide structure of Figure 4.2 and consider the problem as a boundary value problem. Such a problem was discussed in conjunction with Equation (2-15). In fact, by making the assumption that the monochromatic fields are expressible as

$$E(r,t) = Re[E(x,y)e^{i\beta z}e^{-i\omega t}] \tag{4-4}$$

one can express the transverse electromagnetic fields as functions of the normal components in each of the piecewise continuous regions by

$$E_x = \frac{1}{k_i^2 - \beta^2}\left[i\beta\frac{\partial E_z}{\partial x} + i\omega\mu_0\frac{\partial H_z}{\partial y}\right] \tag{a}$$

$$E_y = -\frac{1}{k_i^2 - \beta^2}\left[i\beta\frac{\partial E_z}{\partial y} - i\omega\mu_0\frac{\partial H_z}{\partial x}\right] \tag{b}$$

$$H_x = \frac{1}{k_i^2 - \beta^2}\left[i\omega\epsilon_i\frac{\partial E_z}{\partial y} - i\beta\frac{\partial H_z}{\partial x}\right] \tag{c}$$

$$H_y = -\frac{1}{k_i^2 - \beta^2}\left[i\omega\epsilon_i\frac{\partial E_z}{\partial x} + i\beta\frac{\partial H_z}{\partial y}\right] \tag{d} \quad (4\text{-}5)$$

where the $E_z$ and $H_z$ satisfy the wave equations

$$\frac{\partial^2 H_x}{\partial x^2} + \frac{\partial^2 H_z}{\partial y^2} + (k_i^2 - \beta^2)H_z = 0 \tag{a}$$

$$\frac{\partial^2 E_z}{\partial x^2} + \frac{\partial^2 E_z}{\partial y^2} + (k_i^2 - \beta^2)E_z = 0 \tag{b} \quad (4\text{-}6)$$

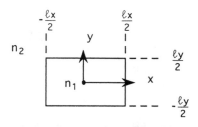

**FIGURE 4-2.**    Rectangular waveguide.

subject to conditions of continuity of all transverse field components across all boundaries. It should be noted that, if the index distribution had been continuously varying rather than piecewise constant, it still would have been possible to derive a system like (4-5), but with index gradients in it. Equations (4-6) would have in that case become coupled second order partial differential equations with nonconstant coefficients.

The partial differential equations of (4-6) look nice, especially when one considers that they can be solved by separation of variables. By assuming that

$$H_z = X_H(x)Y_H(y) \qquad \text{(a)}$$

$$E_z = X_E(x)Y_E(y) \qquad \text{(b)} \quad \text{(4-7)}$$

with

$$\gamma^2 = \gamma_x^2 + \gamma_y^2 = k_i^2 - \beta^2 \qquad \text{(4-8)}$$

one can express the solutions inside the guide as

$$X_E(x) = A_x \cos(\gamma_x x + \phi_x) \qquad \text{(a)}$$

$$Y_E(y) = A_y \cos(\gamma_y y + \phi_y) \qquad \text{(b)}$$

$$X_H(x) = B_x \cos(\gamma_x x + \psi_x) \qquad \text{(c)}$$

$$Y_H(y) = B_y \cos(\gamma_y y + \psi_y) \qquad \text{(d)} \quad \text{(4-9)}$$

The nature of the solutions of the differential equations of (4-6), however, becomes different outside the guide. As illustrated in Figure 4.3, far enough away from the guide's axis, the lateral dimensions of the guide become insignificant, and therefore the guide appears as a line source. Radiation escaping from the guide must therefore die off far enough from the wave-

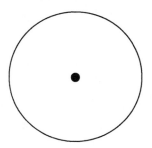

**FIGURE 4.3.**    Far enough away from the axis of the rectangular waveguide, the waveguide appears as a line source, as far as far field radiation is concerned.

guide as an exponential. The solutions to the wave equations in polar coordinates

$$\frac{\partial^2 H_z}{\partial r^2} + \frac{1}{r}\frac{\partial H_z}{\partial r} + \frac{1}{r^2}\frac{\partial^2 H_z}{\partial \theta^2} + (k_i^2 - \beta^2)H_z = 0 \qquad \text{(a)}$$

$$\frac{\partial^2 E_z}{\partial r^2} + \frac{1}{r}\frac{\partial E_z}{\partial r} + \frac{1}{r^2}\frac{\partial^2 E_z}{\partial \theta^2} + (k_i^2 - \beta^2)E_z = 0 \qquad \text{(b)} \qquad \text{(4-10)}$$

which have this "proper" behavior outside the guiding region, would be

$$E_z(r,\theta) = C_v K_v(\kappa r) \cos(v\theta + \phi_v) \qquad \text{(a)}$$

$$H_z(r,\theta) = D_v K_v(\kappa r) \cos(v\theta + \psi_v) \qquad \text{(b)} \qquad \text{(4-11)}$$

where $K_v(\kappa r)$ is a modified Bessel function (Abramowitz and Stegun 1965, Chapter 9) and $\kappa$ is given by

$$\kappa = \sqrt{\beta^2 - k_2^2} \qquad \text{(4-12)}$$

The different nature of the solutions outside and inside can cause a great deal of trouble in obtaining the solution through applying boundary conditions. This should become quite evident to anyone who tries first plugging (4-9) into (4-5), then plugs (4-11) into (4-5), and then tries to equate $E_z$, $H_z$, $E_x$, $H_x$ across the boundaries $y = \pm l_y/2$ and $E_z$, $H_z$, $E_y$, and $H_y$ across the boundaries $x = \pm l_x/2$. The point really is that the functions on the inside of the boundary cannot have the same shape as the functions on the outside of the wall, regardless of the values of the constants $A_x$, $A_y$, $B_x$, $B_y$, $C_v$, $D_v$, $\phi_x$, $\phi_y$, $\psi_x$, $\psi_y$, $\phi_v$, and $\psi_v$. This is because these single functions by themselves cannot form complete sets. In the one-dimensional cases considered in Chapter 2, we never needed complete sets on the boundaries as we needed only one value along the boundary. Here we need to equate functions along complicated boundaries. The solutions in this case will in general have to be infinite sums over complete sets of two-dimensional functions. In fact, Goell (1969) obtained a solution to the rectangular waveguide problem by expanding in infinite sets of Bessel functions both inside and outside the guide and finding a numerical method to equate boundary conditions.

There is really only one electromagnetic boundary value problem that we can exactly solve analytically in two dimensions. Although we will solve this problem in some detail in Chapter 5, we wish to outline its solution here. The idea is to study what can be done exactly in an analytical manner and combine this with some simple approximations to come up with an applicable, understandable approximation procedure. The solvable problem we refer to is, of course, that of the dielectric cylinder, as illustrated in Figure 4.4.

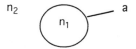

**FIGURE 4.4.**    Dielectric rod problem in which a circular cylinder of index $n_1$ is immersed in a material of index $n_2$.

The important point of this problem is that the two media have the same essential circular symmetry. Equations (4-5) can be transformed to polar coordinates to obtain

$$E_r = \frac{1}{k_i^2 - \beta^2}\left[i\beta\frac{\partial E_z}{\partial r} + \frac{i\omega\mu}{r}\frac{\partial H_z}{\partial \theta}\right] \qquad \text{(a)}$$

$$E_\theta = \frac{1}{k_i^2 - \beta^2}\left[\frac{i\beta}{r}\frac{\partial E_z}{\partial \theta} - i\omega\mu\frac{\partial H_z}{\partial r}\right] \qquad \text{(b)}$$

$$H_r = \frac{1}{k_i^2 - \beta^2}\left[-\frac{i\omega\epsilon_i}{r}\frac{\partial E_z}{\partial \theta} + i\beta\frac{\partial H_z}{\partial r}\right] \qquad \text{(c)}$$

$$H_\theta = \frac{1}{k_i^2 - \beta^2}\left[i\omega\epsilon_i\frac{\partial E_z}{\partial r} + \frac{i\beta}{r}\frac{\partial H_z}{\partial \theta}\right] \qquad \text{(d)} \quad (4\text{-}13)$$

where the $z$-components of the fields satisfy the differential equations of (4-10). Solutions to Equations (4-10) suitable to satisfying the finiteness condition at the origin as well as the outgoing condition at infinity are

$$H_z = \begin{cases} B_v J_v(\gamma r)\sin(v\theta + \psi_{iv}), & r < a \\ D_v K_v(\kappa r)\sin(v\theta + \psi_{ov}), & r > a \end{cases} \qquad \text{(a)}$$

$$E_z = \begin{cases} A_v J_v(\gamma r)\cos(v\theta + \phi_{iv}), & r < a \\ C_v K_v(\kappa r)\cos(v\theta + \phi_{ov}), & r > a \end{cases} \qquad \text{(b)} \quad (4\text{-}14)$$

where $J_v$ is the Bessel function (Abramovitz and Stegun 1965, Chapter 9) and $\gamma$ and $\kappa$ are defined by Equations (4-8) and (4-12), respectively. The sine is chosen in (4-14a) and the cosine in (4-14b) such that each of the fields in (4-13) will have the same $\theta$ dependence in each of its two constituent additive terms. As each transverse field $E_z$, $H_z$, $E_\theta$, and $H_\theta$ has the same $\theta$ dependence inside and out, by setting the phase functions $\psi_{iv}$, $\psi_{ov}$, $\phi_{iv}$, and $\phi_{ov}$ equal, the azimuthal parts of the boundary conditions are automatically satisfied. What is left is only to equate the radial parts of the transverse fields $E_z$, $H_z$, $E_\theta$, and $H_\theta$ across the boundary at $r = a$. This amounts to solving the radial equations

$$\frac{\partial^2 H_z}{\partial r^2} + \frac{1}{r}\frac{\partial H_z}{\partial r} + \left(k_i^2 - \beta^2 - \frac{v^2}{r^2}\right) H_z = 0 \qquad \text{(a)}$$

$$\frac{\partial^2 E_z}{\partial r^2} + \frac{1}{r}\frac{\partial E_z}{\partial r} + \left(k_i^2 - \beta^2 - \frac{v^2}{r^2}\right) E_z = 0 \qquad \text{(b)} \quad (4\text{-}15)$$

which have the solutions

$$H_z = \begin{cases} B_v J_v(\gamma r), & r < a \\ D_v K_v(\kappa r), & r > a \end{cases} \qquad \text{(a)}$$

$$E_z = \begin{cases} A_v J_v(\gamma r), & r < a \\ C_v K_v(\kappa r), & r > a \end{cases} \qquad \text{(b)} \quad (4\text{-}16)$$

subject to boundary conditions of continuity of $E_z$, $E_\theta$, $\epsilon_i H_z$, and $\epsilon_i H_\theta$. What results is four equations in four unknowns, $A_v$, $B_v$, $C_v$, and $D_v$. Just as in the one-dimensional cases discussed in Chapter 2, the system of equations is homogeneous, and therefore, a determinantal equation determines the propagation constant $\beta$ and back substitution leads to expressions for the unknowns as a one-parameter string. The reason why the solution is so simple relates to the separability of the boundary conditions or, more simply stated, as in the one-dimensional case of Chapter 2, the functions of (4-16) have only one value at the boundary $r = a$.

That it is in general necessary to use an expansion in an infinite set of complete modes inside each region of a piecewise constant index boundary value problem is a great barrier to analytical study of such structures. It should not really be a barrier to understanding of such structures, however. For example, a small noncircular perturbation to a circular waveguide will render the problem analytically intractable yet should not qualitatively alter the mode structure in any significant way (excepting the cases of broken degeneracies). The point is that we do not have the adequate mathematical tools. In the last two centuries, it was common practice for mathematicians when faced with differential equations they could not solve to say that the differential equation defined a new kind of function, of which other mathematicians would then start cataloging the properties. The function would generally end up with the name of the mathematician first identified with it. A beautiful example of such a case is given by the treatise (804 pages long) on the Bessel function by Watson (1966). The most recent case of such a function definition that I know of is by Maliuzhinetz (1958). However, this was one of the few cases of the cataloging of a function that I know of in this century. The reasons for this are manyfold. An important one is that these days, engineers and physicists come up with so many unsolvable differential equations that there just are not enough mathematicians to go around to catalogue all the possible solution functions. More recently, we have entered an era where computer power is so cheap and easily obtainable that it is much easier to grind out the solution to a differential equation than it is to look up its properties. In trying to come to an understanding of waveguide properties, however, what is necessary is not a detailed knowledge of mathematical properties of the waveguide solutions nor high accuracy numerical solutions to mathematical models of which the parameters are poorly known, but rather a qualitative picture of the shape of the modes and some ideas about resonant frequencies, losses, etc.

We now wish to look briefly at a solution to the rectangular waveguide problem of Figure 4.5 which was originally given by Marcatilli (1969). The philosophy of the solution is that, for a guided field, there must be much higher field intensity in the core than in the cladding. Therefore, rather than be tied to a condition at infinity as an exact solution to the problem would be, why not tie the solutions at infinity to the separable solution in the core? A possible way to do this would be to essentially split the problem up into two problems, one along the $x$ axis and one along the $y$ axis. With this, one could write that

$$H_x = \begin{cases} M_1 \cos(\gamma_x x + \phi) \cos(\gamma_y y + \psi) & \text{in I} \\ M_2 \cos(\gamma_x x + \phi) \, e^{-\kappa_{y2} y} & \text{in II} \\ M_3 e^{-\kappa_{x3} x} \cos(\gamma_y y + \psi) & \text{in III} \\ M_4 \cos(\gamma_x x + \phi) \, e^{+\kappa_{y4} y} & \text{in IV} \\ M_5 e^{+\kappa_{y5} x} \cos(\gamma_y y + \psi) & \text{in V} \end{cases} \tag{4-17}$$

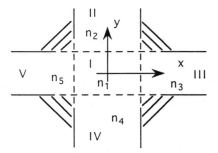

**FIGURE 4.5.**    Rectangular waveguide problem in which one chooses to ignore certain regions in which it is hard to obtain a solution.

or that

$$
E_x = \begin{cases}
M_1 \cos(\gamma_x x + \phi) \cos(\gamma_y y + \psi) & \text{in I} \\
M_2 \cos(\gamma_x x + \phi)\, e^{-\kappa_{y2} y} & \text{in II} \\
M_3 e^{-\kappa_{x3} x} \cos(\gamma_y y + \psi) & \text{in III} \\
M_4 \cos(\gamma_x x + \phi) e^{+\kappa_{y4} y} & \text{in IV} \\
M_5 e^{+\kappa_{x5} x} \cos(\gamma_y y + \psi) & \text{in V}
\end{cases}
\qquad (4\text{-}18)
$$

where

$$\gamma_x = \sqrt{k_i^2 - \beta_x^2} \qquad (a)$$

$$\gamma_y = \sqrt{k_i^2 - \beta_y^2} \qquad (b)$$

$$\kappa_{xi} = \sqrt{\beta_x^2 - k_i^2} \qquad (c)$$

$$\kappa_{yi} = \sqrt{\beta_y^2 - k_i^2} \qquad (d) \quad (4\text{-}19)$$

and $M_1$, $M_2$, $M_3$, $M_4$, $M_5$, $\phi$, and $\psi$ are parameters to be determined from boundary conditions. Implicit in this form of solution is the assumption that the modes are linearly polarized and plane-wave-like in nature and, therefore, that either (4-17) or (4-18) are sufficient to solve the problem at hand. Specifically, to solve for the so-called $E_{ypq}$ (where the y denotes linear polarization direction and the $p$ and $q$ are the mode orders in the $x$ and $y$ directions respectively) modes, one would use the $H_x$ of (4-17) to satisfy the boundary conditions on $H_x$ and $H_z$ at the I-II and I-IV interfaces. One could then use the associated $E_y$ field to satisfy the boundary conditions on $E_y$ and $H_z$ at the I-III and I-V interfaces. The situation here is analogous to that in Chapter 2, Equations (2-16)–(2-22), except that here we need to use $H_x$, $\partial H_x/\partial x$ (see Equations (2-16b) and (2-16c)) for the tangential $H$ field on interfaces I-II and I-IV but $E_y$ and $\partial E_y/\partial x$ (again Equations (2-16b) and (2-16c)) for an $E$ tangential and an $H$ tangential on I-III and I-V. As there are four constants, $M_1$, $M_2$, $M_3$, and $\phi$, associated with the four boundary conditions on I-II and I-IV, these equations can be used to determine $\gamma_y$ and $M_3$, $M_5$, and $\psi$ as functions of $M_1$. The equations on I-III and I-V can likewise be used to determine $\gamma_x$ and $M_2$, $M_4$ and $\phi$ as functions of $M_1$. As in Chapter 2, equations (2-51)–(2-58), the $M_1$ can then be used to normalize the modal wave function. A small problem occurs here, though, as there are two values of $\beta$, $\beta_x$ and $\beta_y$. This problem will be taken up in the next paragraph. The important point is that all the steps necessary to define a linearly polarized mode have been carried out analytically with the exception of a numerical determination of $\beta$ values. Further, it is clear that an analogous procedure could be carried out to determine the $H_{ynm}$ modes, which would be primarily $x$-polarized.

Before turning the discussion to what is wrong with this type of solution, perhaps what is right with this solution should first be discussed. As was mentioned, the solutions could be denoted as $E_{ymn}$ and $H_{ymn}$. The $m$ and the $n$ should denote the number of zeroes occurring along the $x$ and $y$ axes, respectively. Evidently, two-dimensional intensity plots of the optical intensity in the core region for the first few $E_{ymn}$ modes should appear as in the sketches of Figure 4.6, where high intensity is denoted by darker color. It should be noted that the intensity plots of Marcatilli's work closely resemble those of Goell, even though cases can be found where the $\beta$'s differ by some amount. Clearly, Marcatilli's solution is at least qualitatively correct.

There are clearly problems with Marcatilli's solution. That there are two distinct values of $\beta$ is not good, as the mode cannot be normalized and phase velocities, etc., are double-valued. To predict dispersion characteristics, an important prediction for an analytical theory as it is quite costly to carry out numerically, something must be put in by hand. This is perhaps the most important failing of the method. Other problems are perhaps more cosmetic. For example, the fields cannot be continuous at the waveguide corners or along infinite curves extending from the corners to infinity in the regions that were removed from active consideration in the problem. The fields at the corners may have some appreciable value, but this rapidly decays in the direction of the discontinuity. Further, it is dispersion and loss that really determine waveguiding properties. The details of the field distribution effect overlap integrals of coupled mode calculations (see Chapter 5), but then the errors are under integrals and get averaged. That the fields at infinity do not die uniformly as they should in exact theory is perhaps the least of our worries, as the fields at infinity truly are small. In conclusion, one could say that, if one could find a fix-up for the two-$\beta$ dilemma of Marcatilli, one might have the basis for a nice analytical approximation to the rectangular waveguide problem.

A more systematic way to go about solving such a boundary value problem as the rectangular waveguide is embodied in the effective index method (Knox and Toulios 1970). The idea behind this technique can be depicted as in Figure 4.7. The idea here is that one does not know how to solve a

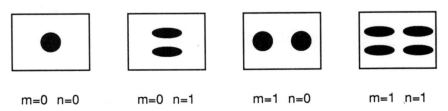

m=0  n=0          m=0  n=1          m=1  n=0          m=1  n=1

**FIGURE 4.6.**    Two-dimensional intensity plots of the first few $E_{ymn}$ modes of a rectangular dielectric waveguide.

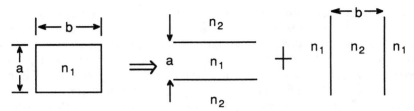

**FIGURE 4.7.**    Schematic depiction of the process behind the effective index method.

rectangular waveguide problem but one does know how to solve essentially any slab problem. The procedure then is to break up the unsolvable problems into simpler, solvable ones. As all trial solutions such as those of (4-17) and (4-18) will be parameterized by $\beta$, we first and foremost need to use a technique that can get us close to a correct $\beta$ value. One could, for example, consider the problem of Figure 4.8 as defining a value of $\beta$ that is a function of the guide width $b$. The curve $\beta(b)$ may look like that sketched in Figure 4.8. The idea here is that, in the limit where $b$ goes to zero, there is no guiding layer and the wave must just propagate as a plane wave in a medium of index $n_2$. In the case where $b$ becomes infinite, then the problem reduces to a slab problem in which, although we cannot write down a $\beta$ value, we know how to obtain one, and we further know that it will lie between $n_2 k_0$ and $n_1 k_0$. In the situation where we consider a well bound fundamental mode, the $b$ value should lie considerably closer to $n_1 k_0$ than to $n_2 k_0$.

One can then consider the problem of Figure 4.9. Here we consider horizontally segmenting the two-dimensional problem into three one-dimensional slab problems. These three problems represent the limiting cases of the plot of Figure 4.9. With reference to the sketch of 4.8, let us make

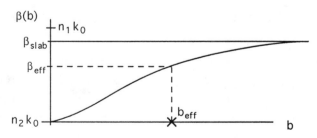

**FIGURE 4.8.**    Possible variation of $\beta$ with waveguide width $b$ for a rectangular dielectric waveguide.

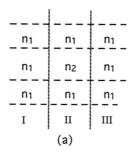

(a)

Problem I                    Problem II                    Problem III

$n_1$

n₁                            n₂                            n₁

$n_1$

(b)                          (c)                          (d)

**FIGURE 4.9.**    Possible segmentation of a dielectric waveguide problem where (a) depicts the initial problem with horizontal regions I, II, and III labeled and (b), (c), and (d) split off the I, II, and III regions from the structure for separate consideration.

the identifications that

$$n_{\mathrm{I}} = \frac{\beta(0)}{k_0} = n_{\mathrm{III}}$$     (a)

$$n_{\mathrm{II}} = \frac{\beta(\infty)}{k_0}$$     (b)   (4-20)

Now we wish to consider a horizontal problem such as Figure 4.10. Here the "effective" indices of each of the three problems of Figure 4.9(b–d) have been used to replace the horizontal layers I, II, and III of Figure 4.9(a). We have now replaced the index distribution of the central region with a single effective index that was derived from the slab problem of Figure 4.9(c) above.

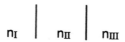

$n_{\mathrm{I}}$        $n_{\mathrm{II}}$        $n_{\mathrm{III}}$

**FIGURE 4.10.**    An illustration of the fourth problem associated with an effective index solution to the problem of Figure 4.9(a).

We know that the index we find for the problem of Figure 4.10 will be bounded between $n_2 k_0$ and $n_{II} k_0$. This effective $\beta$ value $\beta_{eff}$ will correspond on Figure 4.8 to some effective width $b_{eff}$, as marked in Figure 4.8. Although it is not at all clear how close this $b_{eff}$ of Figure 4.8 will be to the actual $b$ of Figure 4.7, it is clear that the correction to the $\beta$ value that comes from solving the problem of Figure 4.10 will put one closer to the true $b$ value. The effective index $n_{eff}$, defined by

$$n_{eff} = \frac{\beta_{eff}}{k_0} \tag{4-21}$$

will correspond to the $\beta_{eff}$ that will be used as the mode propagation constant.

To get the fields for the rectangular guide problem of Figure 4.9(a), one can use the following effective index procedure. We see that we must solve four problems, those of Figures 4.9(b–d) and 4.10, which we will label problems I, II, III, and IV, respectively. Clearly, problems I and III are already solved, as we already know the $\beta$ values in these regions. As we shall soon see, we do not need the fields in these regions. Problem II must be solved, with the resulting solution

$$\psi_{II}(y,\beta_{II}) = \begin{cases} A_{II+}e^{-\kappa_y y}, & y > ly/2 \\ A_{II}\cos \gamma_y y + B_{II}\sin \gamma_y y, & -ly/2 < y < ly/2 \\ A_{II-}e^{\kappa_y y}, & y < -ly/2 \end{cases} \tag{4-22}$$

where the $\beta_{II}$ value is obtained from setting the $T_{11}$ element of the boundary condition matrix equal to zero (see the development of Chapter 2). The dependence of the parameters of this solution on $\beta_{II}$ is explicitly noted on the left-hand side of (4-22). One can also solve the problem of IV, with the result

$$\psi_{IV}(x,\beta_{IV}) = \begin{cases} A_{IV+}e^{-\kappa_x x}, & x > lx/2 \\ A_{IV}\cos \gamma_x x + B_{IV}\sin \gamma_x x, & -lx/2 < x < lx/2 \\ A_{IV-}e^{\kappa_x x}, & x < -lx/2 \end{cases} \tag{4-23}$$

The solution to the two-dimensional problem, which we denote $\psi_{eff}(x,y,\beta_{IV})$, is given by

$$\psi_{eff}(x,y,\beta_{IV}) = \psi_{II}(y,\beta_{IV})\psi_{IV}(x,\beta_{IV}) \tag{4-24}$$

where the $\beta_{II}$ in the solution of (4-22) has been replaced by $\beta_{IV}$. This replacement allows us to now talk about modal normalization, modal dispersion, etc., as we can define a single modal propagation constant. It also is the

| $e^{\kappa_x x} e^{-\kappa_y y}$ | $\cos\gamma_x x \; e^{-\kappa_y y}$ | $e^{-\kappa_x x} e^{-\kappa_y y}$ |
|---|---|---|
| $e^{\kappa_x x} \cos\gamma_y y$ | $\cos\gamma_x x \cos\gamma_y y$ | $e^{-\kappa_x x} \cos\gamma_y y$ |
| $e^{\kappa_x x} e^{\kappa_y y}$ | $\cos\gamma_x x \; e^{\kappa_y y}$ | $e^{-\kappa_x x} e^{\kappa_y y}$ |

$$-\ell_x/2 \qquad\qquad \ell_x/2$$

**FIGURE 4.11.** Schematic depiction of the forms of the effective index solution in all of the nine regions defined by the problem.

source of error in the solution technique. Equation (4-22) with $\beta_{\mathrm{II}}$ exactly solves the one-dimensional boundary conditions. With $\beta_{\mathrm{IV}}$, it only approximately solves the boundary conditions. However, this approximate solution is probably about as good as can be obtained simply. Had the index profile been continuous, one could have chosen the $\psi(y)$ which corresponds most closely to the obtained $\beta$. This may be a better solution than $\psi_{\mathrm{II}}$.

The solution of (4-24) can be schematically depicted as in Figure 4.11. In the figure, the sinusoidal solution of the form $A \sin + B \cos$ has simply been replaced by cos to leave the diagram less cluttered. As is immediately evident from the diagram, the effective index method has obtained solutions in the corner regions which were ignored by Marcatilli. However, the discontinuities and behavior at infinity are really no better than in Marcatilli's solution. The thing that has changed is that a single $\beta$ has been defined for the problem and, further, a uniform procedure for obtaining that $\beta$ and the corresponding fields has been obtained. An interesting point to note is that the assumption of a single $\beta$ has caused the $\gamma_x$ and $\gamma_y$ values to become identical, and as with $\kappa_x$ and $\kappa_y$, as they should be.

The effective index procedure can be applied to any index distribution. Consider the distribution sketched in Figure 4.12. What is done is first to inscribe the index distribution in a rectangle whose sides lie outside the

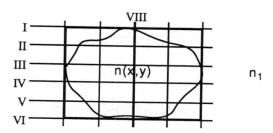

**FIGURE 4.12.** Generic bounded index distribution.

varying index region. (If the varying index region extends to infinity, the techniques for applying the effective index method can be found. An example is given in the last section of this chapter.) One then discretizes the medium within the rectangle and forms a "staircase" of constant index media. Referring to Figure 4.12, it is clear that problems I and VI are the trivial ones, and the others, problems II–V, must be solved by some technique like the matrix one of Chapter 2. Once all the horizontal problems are solved, then one can replace the two-dimensional index distribution by a one-dimensional staircase index distribution $n_{\text{eff}}(y)$. Problem VII can then be solved, and an effective index $n_{\text{eff}} = \beta_{\text{eff}}/k_0$ can be defined for the waveguide, whose field distribution can be found analogously to the rectangular guide distribution found above.

Some comments on the mathematical validity of the effective index technique should perhaps be made at this point. What one actually is trying to solve in the general (weakly guiding) case, is an equation of the form

$$\nabla_t^2 \psi(x,y) + (k^2(x,y) - \beta^2)\psi(x,y) = 0 \qquad (4\text{-}25)$$

where $\nabla_t^2$ is the Laplacian in the transverse $(x,y)$ coordinates, and $k^2(x,y)$ is some index function. Now, formally at least, one can always try separation of variables. Assume

$$\psi(x,y) = X(x)Y(y) \qquad (4\text{-}26)$$

and plug (4-26) into (4-25), to find

$$\frac{1}{Y}\frac{\partial^2 Y}{\partial y^2} + \frac{1}{X}\frac{\partial^2 X}{\partial x^2} + (k^2(x,y) - \beta^2) = 0 \qquad (4\text{-}27)$$

which can be separated into

$$\frac{\partial^2 Y}{\partial y^2} + k_y^2(x,y)Y = 0 \qquad \text{(a)}$$

$$\frac{\partial^2 X}{\partial x^2} + k_x^2(x,y)Y = 0 \qquad \text{(b)} \quad (4\text{-}28)$$

where

$$k_x^2(x,y) + k_y^2(x,y) = k^2(x,y) - \beta^2 \qquad (4\text{-}29)$$

Now, obviously, Equations (4-28) are nonsense as they are unsolvable, that is to say, they did not separate. However, consider the case in which $k^2(x,y)$ is slowly varying in $y$, which is often notated as

$$k^2(x,y) = k^2(x;y) \tag{4-30}$$

which really means that $k^2(x,y)$ is sufficiently slowly varying in $y$ that $y$ can be considered as a parameter and not a variable. Let

$$k_x^2(x;y) = k^2(x;y) - k_y^2(y) \tag{a}$$

$$k_y^2(y) = \beta_x^2(y) - \beta^2 \tag{b} \quad (4\text{-}31)$$

Then Equation (4-29) is satisfied, and the system

$$\frac{\partial^2 X(x;y)}{\partial x^2} + (k^2(x;y) - \beta^2(y))X(x;y) = 0 \tag{a}$$

$$\frac{\partial^2 Y(y)}{\partial x^2} + (\beta_x^2(y) - \beta^2)Y(y) = 0 \tag{b} \quad (4\text{-}32)$$

is approximately separated. The error involved in carrying out such a procedure is still a matter of research.

Now the above outlined procedure is one for finding the $\beta$. It is not necessarily true that using the relation

$$\psi(x,y) = X(x;y)Y(y) \tag{4-33}$$

as in Equation (4-24) is the best way to determine the field structure. As was mentioned previously, the change in $\beta$ in the $X(x;y)$ can cause serious errors in the field structure. How to handle this problem is a hotly contested issue. One suggested way which has found some success in practice is to back substitute the $\beta_x(y)$ and $\beta$ into the $X(x;y)$ equation to obtain the modified equation

$$\tilde{X}(x;y) + X(\beta_x^2(y) - \beta^2)\tilde{X}(x;y) = 0 \tag{4-34}$$

and to obtain the fields as

$$\psi(x,y) = \tilde{X}(x;y)Y(y) \tag{4-35}$$

It is clear now that, for any two-dimensional index distribution which we would want to stick in a cavity, we can define cavity modes and the $\mathbf{u}_\lambda(\mathbf{r})$ and $\omega_\lambda$ of Equation (4-1). We shall look at a realistic example of such a guide in the last section of this chapter. Of course, one of the main points of all that we have done in the last chapter and a half is to include the index of refraction of a dynamical variable through the polarizability $\alpha$ and the inversion $d$. To define a constant index distribution $n(x,y)$ is a little bit of a strange thing to do, but not really so strange. In all that we have done here, we have considered a two-level system. No material system consists of only two levels. There are a large number of resonances at various frequencies, and all of the resonances contribute to the index of refraction. Although the band gap of a semiconductor is an important resonance, it does not change the total index of refraction as much as one percent. We can therefore talk about a large background index and a small dynamic index. It is the large background index that goes into the calculation of $u_\lambda(r)$ and $\omega_\lambda$. The dynamic index will dynamically perturb these quantities according to the predictions of the Maxwell-Bloch system. It is for this reason that $E_\lambda(t)$ varies with time. Were there no dynamic index and were the resonator walls really 100% reflecting, the resonances would be perfect and the $E_\lambda$ would vary exactly as $e^{-i\omega_\lambda t}$. Between the dynamic index and the cavity leakage, these resonances get smeared out and laser dynamics result.

The only field parameter we have yet to determine is $\kappa_\lambda$, the loss parameter. We know how to relate this parameter to $\sigma$, but as was discussed at some length in Chapter 2, we have no real concept of what $\sigma$ is (or perhaps too many concepts of what $\sigma$ is). Therefore we shall use an empirical argument to determine $\kappa_\lambda$. The point here is that, in a well constructed laser cavity, the main source of loss should be leakage through the mirrors. This should especially be true in a semiconductor laser where the mirror reflectivity is low with respect to unity. The situation is as illustrated in Figure 4.13. Light bounces back and forth between the mirrors, with some portion being leaked on each pass. The $\kappa_\lambda$ is essentially the inverse of the photon lifetime, which is the time that some intensity of light inside the empty cavity would take to decay to its $1/e$ point. The question is, therefore, how many trips $N$ does light take between the mirrors before this decay? If leakage is the major loss

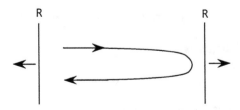

**FIGURE 4.13.**     Leakage of light from a Fabry-Perot resonator with mirrors of finite reflectivity $R$.

mechanism, this $N$ would be defined by

$$R^N = 1/e \tag{4-36}$$

Equation (4-23) can be solved as

$$N = \frac{1}{\log R} \tag{4-37}$$

In laser cavities with high finesse—that is, $R \sim 1$—one generally writes

$$\log R = -\log \frac{1}{R} = -\log \left(1 + \frac{1-R}{R}\right) \tag{4-38}$$

and expands the logarithm to obtain

$$N = \frac{R}{1-R} \tag{4-39}$$

In GaAs, the reflection is due to the index mismatch between the GaAs index of 3.2 and the index of air, which is roughly one. The $R$ defined by this process is therefore

$$R = \left(\frac{3.2 - 1}{3.2 + 1}\right)^2 = 0.3 \tag{4-40}$$

which is not close to 1. The point is that, using either (4-36) or (4-39), one obtains the result that light makes about one pass. This means that on the average a photon radiated in the cavity just goes right out of it. This is a prime reason why semiconductor lasers can be modulated at such high rates. It is also a cause of high threshold current. At any rate, to turn this number of passes into a time, one needs to consider the cavity transit time. A semiconductor laser is generally on the order of 300 $\mu$m long. The speed of light in the material is given by $c/n$. This is not quite the speed we want, however, as we really want the speed of energy transport, which is given by $c/n_g$, where $n_g$ is the group velocity. For GaAs, around the band gap this group index is about 4. Putting this together, we should get the cavity transit time, which is just the photon lifetime $\tau_p$, given by

$$\tau_p = \frac{n_g l}{c} = 4 \text{ psec} = \frac{1}{\kappa_\lambda} \tag{4-41}$$

## 4.2 BAND STRUCTURE

Now that all the field parameters are well defined, it is time to define the material parameters. To do this, a very simple model of band structure (see, for example, Kittel 1971, Chapters 9 – 10) will be presented for the purpose of understanding the relation between the semiconductor and the two-level atomic system we considered in the derivation. First, consider an electron in free space whose wave function may be given by the plane wave

$$\psi_e(\mathbf{r},t) = Ne^{i\mathbf{k}\cdot\mathbf{r}}e^{-i\frac{E}{\hbar}t} \tag{4-42}$$

where $N$ is a normalization constant and the dispersion relation between the wave vector $k = p/\hbar$ and the energy $E$ is given by

$$E = \frac{\hbar^2 k^2}{2m} \tag{4-43}$$

which is plotted in Figure 4.14. One could consider the case of an "almost" free electron, as sketched in Figure 4.15. The idea here is that, if a higher valence impurity is present in the lattice, at least one electron will not be involved in the covalent bonding of the lattice. This electron will only be electrostatically bound to the lattice site. As electrostatic bonding is much

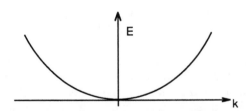

**FIGURE 4.14.**    Dispersion relation for a free electron.

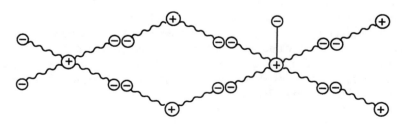

**FIGURE 4.15.**    Part of a lattice which contains an impurity with a higher valence than the 4 valence of the primary lattice constituent.

weaker than covalent bonding, it takes very little energy to free this electron into the material. If such impurity sites are rare in the lattice, the electron could wander through the lattice for a long time before sticking to another site. It is just such electrons that we want to consider in our model.

What does our almost free electron see as it propagates through the lattice? It will basically see a periodic potential such as the one sketched in Figure 4.16. The point here is that the electron will feel attraction to the lattice sites with the excess positive charge, yet repulsion by the net negative charge of the electron bonding clouds. A major element of the theory of wave propagation in a periodic potential is the concept of Bragg reflection, as illustrated in Figure 4.17. The planes in the figure represent the atomic planes, and the arrowed lines represent the directions of plane wave propagation. It is readily shown that, for normal incidence on such a stack of planes, the reflected waves will interfere constructively when $2kd = 2n\pi$ or for values of the $k$ vector

$$k = \frac{n\pi}{d}$$

(4-44)

In optics, when partial reflections add up in phase, this indicates that a

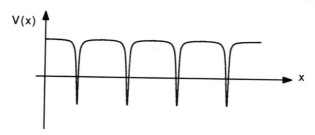

**FIGURE 4.16.**    A periodic potential as might be seen by an electron propagating through a lattice.

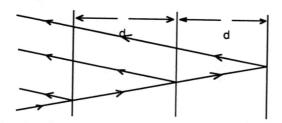

**FIGURE 4.17.**    The process which gives rise to Bragg reflection.

wave will be strongly reflected from a structure. In the case we are considering here, however, the electron is already inside the material. It can not come flying out of the material because of the material's surface potential, and if it reflects it will just reflect again, because the material looks the same in the backward direction as it does in the forward direction. The point here is that the electron can only exist as a standing wave for these $k$ values. There are two possible standing waves, however, one with maxima at lattice sites and one with minima at lattice sites. As the lattice sites carry positive charge, the standing wave with maxima on these sites will be a lower energy solution than the other. This splitting of states gives rise to a bandgap between the lower band's upper standing wave state and the upper band's lower standing wave state. The effect this has on the "almost" free electron dispersion relation (as opposed to the free electron dispersion relation of Figure 4.14) is sketched in Figure 4.18. Note that small circles have been drawn in on the diagram to indicate that the available $k$-states are actually quantized. That is, since the crystal (or domain) size must actually be finite, there is a minimum value which the $k$ vector can take on, as illustrated in Figure 4.19. This minimum $k$ value will also be the $k$ increment, at least in the limit where one considers the box to be an infinite square well. As the $d$ value is of the order of a few angstroms and the crystal size is minimally of a few hundred microns in our case, the minimum $k$ spacing would be about $10^{-5}$ times the band width (i.e., distance between successive gaps), or there would be minimally about $10^{15}$ states per three-dimensional band. This seems to be sufficient to consider the band continuous, but we shall soon see that due to Fermi statistics the quantum nature of the states will become all-important in laser operating characteristics.

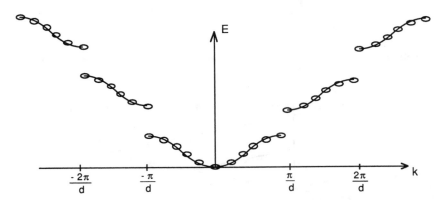

**FIGURE 4.18.**    Dispersion relation of a particle moving in a periodic medium, where the circles on the branches of the function are simply to indicate that the states must be quantized.

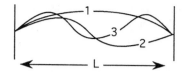

**FIGURE 4.19.**    Illustration of the solution for the first few low energy states of a particle in a box.

Other than the little open circles, the important fact in Figure 4.18 is the existence of gaps that do not show up in Figure 4.14. These gaps have greater impact on electron behavior than simply being forbidden energy regions. As can be seen from Figure 4.18, the curvature of the $\epsilon$-$k$ diagram is reversed below and above the gap. The meaning of this can be ascertained from the following argument. Let us say that the force on a particle is given by

$$F = \frac{dp}{dt} \tag{4-45}$$

in general. In quantum mechanics, $p$ is given by $p = \hbar k$, whereas the motion of a classical point mass is given by $p = mv$. We would like to find the effective value of the electron mass, m*, in the potential of Figure 4.16. This cannot be the same as that of the electron rest mass, as the wave function of the electron will be seriously affected by the potential, at least in the neighborhood of the gaps, and therefore the point mass form of the force law cannot apply. One needs to correct this form. Let us begin by assuming that one can use the point mass force law if one replaces $m$ by the effective mass m* and $v$, the particle velocity, by the group velocity $v_g$ of the particle's wave function. One could therefore write that

$$F = m \frac{dv_g}{dt} \tag{4-46}$$

In wave mechanics, one can always express the group velocity as $d\omega/dk$ (standard dispersion theory; see Brillouin 1960, for example) to obtain

$$F = m* \frac{d}{dt} \frac{d\omega}{dk} = m* \frac{d^2\omega}{dk^2} \frac{dk}{dt} \tag{4-47}$$

Taking the particle's (wave's) energy $\epsilon$ to be $\hbar\omega$, one obtains

$$F = \frac{m*}{\hbar} \frac{\partial^2 \epsilon}{\partial k^2} \frac{\partial k}{\partial t} \tag{4-48}$$

Now, this expression should be identical with (4-45) if $p$ is replaced by $\hbar k$, the quantum mechanical relation, yielding

$$\frac{m^*}{\hbar^2} \frac{\partial^2 \epsilon}{\partial k^2} = 1 \qquad (4\text{-}49)$$

or

$$m^* = \frac{\hbar^2}{\partial^2 \epsilon / dk^2} \qquad (4\text{-}50)$$

Equation (4-50) has some important consequences. An important one is that it is correct in the free particle limit. That is, if one were to plug (4-42) into (4-50), one would obtain $m^* = m$, a very heartening result, as in the free particle case there is no potential to distort the electron wavefunction's inner workings and therefore the electron's force law should be like that of point mass. Directly above the first bandgap (that is $\pi/d$) in Figure 4.18, the behavior is similar but not identical to this. The point is that $\partial^2 \epsilon / \partial k^2$ is positive, but its detailed value depends on all the material parameters. Therefore, the effective mass could be greater or less than the "bare" mass. Below the gap, the situation is strikingly different. Here, $\partial^2 \epsilon / \partial k^2$ is negative, and therefore the effective mass is negative. If the particle is given a push in a given direction, it goes the wrong way, with the reversal energy being supplied by the lattice. This situation is explainable if one thinks of the forms of the wave functions directly above and below the energy gap. As the electron energy approaches the gap energy, more and more of the probability density is reflected, until at the gap the wave function becomes a standing wave. As a zero phase point is defined by the lattice sites, there are two possible standing wave solutions, as are illustrated in Figure 4.20. From these, the force response behavior should become clear. Figure 4.20(a) illustrates a negative

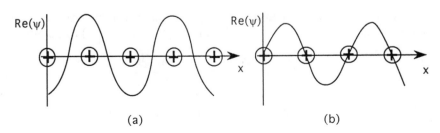

FIGURE 4.20.    Two independent standing wave functions: (a) a possible valence wavefunction, (b) a possible conduction band wavefunction.

phase charge cloud wave function which is most likely to be sitting on a positively charged lattice point. This corresponds to a minimum energy state for a standing wave and therefore corresponds to a valence electron. Trying to push this wave function off this lattice point will lead to a restoring force and will therefore cause the "particle" to move the wrong direction with applied force. Figure 4.20(b) illustrates a conduction band wave function which is in unstable equilibrium (that is, unstable if the state of 4.20(a) is unoccupied and 4.20(b) can collapse into it). A push of this particle will cause motion along the direction of the applied force.

Before incorporating the knowledge we have just obtained into the semiconductor form of the semiclassical laser equations, perhaps we should plot the band structure of a semiconductor in its "standard" form. The idea here is that there is a symmetry in $2\pi/d$. For example, one considers a particle in the lowest band of Figure 4.18 in a state right at the edge $\pi/d$, as we saw previously. If one applies a push to the right on the particle, it will suddenly reappear at $-\pi/d$ because it wants to go in the wrong direction. One could also try this in the next higher band of Figure 4.18 to convince oneself that the points $2\pi/d$, 0, and $-2\pi/d$ in this band are equivalent, as were the values $\pi/d$ and $-\pi/d$ in the preceding example. This should convince one that bands can be translated by $2\pi/d$ and, therefore, that one can apply $2\pi/d$ translations to the second and third bands of Figure 4.18 to obtain the reduced band scheme of Figure 4.21. In this scheme, it is clear that the lower band can be identified with the valence band and that the upper band can be identified with the conduction band. It is also clear that this is a "principal value" scheme in that one need not worry about the $2\pi/d$ translations that the lattice can impart to the wave vector. In the next section, we will use knowledge of this energy wave vector relationship to rewrite the laser equations of Chapter 3 in a semiconductor laser form where electrons will be labeled by the **k** vector rather than by their atomic coordinate $x_\mu$. Note, however, that in the majority of the discussion given here, a 1-dimensional representation is taken as being sufficient, and therefore **k** is generally denoted only by its scalar magnitude $k$ and only when necessary by its vector representation **k**.

## 4.3 FERMI LEVELS

A very important point about electrons in a lattice is that they must obey Fermi statistics. The basis for these statistics is the fact that no two electrons can occupy the same state. This condition was certainly obeyed in the atomic case, where we assumed that each atom had a single polarizable electron. We gave no consideration to cases in which more than one electron were on a single atom. As the atoms were distinct, electrons on different atoms could

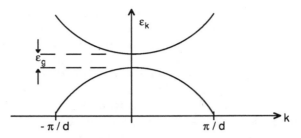

**FIGURE 4.21.**    Reduced band scheme for the extended band structure of Figure 4.18.

have similar states, but they were not identical as they were spatially distinct. In the semiconductor case, the electrons are not localized, and therefore they all occupy the same spatial region. Now electrons can have two spin states, implying that two oppositely spinning electrons can occupy the same state. For the moment, we will ignore this spin degeneracy. In light of this for the electron states to be different, therefore, the condition that no two identically spinning electrons can occupy the same $k$ state must be obeyed, where the $k$ states are as they were depicted in Figure 4.18 and discussed in the paragraph accompanying the figure. This condition tells us little about the band filling, as the electrons can still be distributed in any given manner as long as no two electrons sit in the same $k$ state. In the atomic case, the condition of thermal equilibrium told us that each and every electron, regardless of position, had a given probability to be in its upper state and a given probability to be in its lower state. These probabilities could be used to calculate an average electron energy over the ensemble. As it turns out, the situation, in terms of energy at least, is equivalent here, and $f(\epsilon)$, the average occupancy of a state with energy $\epsilon$, is given by

$$f(\epsilon) = \frac{1}{1 + \exp[(\epsilon - \mu)/k_B T]} \qquad (4\text{-}51)$$

which is plotted for several values of $T$, the temperature, in Figure 4.22. A main point to be seen from the plots is that the weakly temperature dependent Fermi-level $\mu$ serves as the average value of the energy in the sense that it is the value at which the distribution function goes through one-half. In combination with the Fermi exclusion principle, the Fermi level determines a material's electrical properties. For example, consider the illustrations of Figure 4.23. The most important electrical property of a material is its conductivity, which is determined by the number of carriers available for

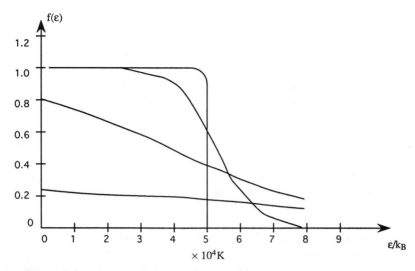

**FIGURE 4.22.**   Fermi distribution function for several values of the temperature.

conduction current. In a semiconductor or insulator, the Fermi level lies within the energy gap. Depending on whether $k_B T$ (where $k_B$ is Boltzman's constant) is greater or less than the energy separation of the Fermi level and conduction band edge, there will be few or no carriers available. By contrast, in a conductor, the Fermi level lies within the conduction band, and therefore there are many current carriers available at any temperature.

The nice parabolic form of the bands of Figure 4.23 is not the only form that bands can take. Examples of an indirect gap semiconductor and a semimetal are sketched in Figure 4.24. As we shall see in the next paragraph, however, these band structures are of no interest to us here as these structures

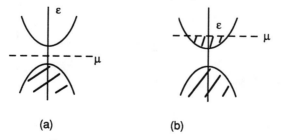

**FIGURE 4.23.**   Positioning of the Fermi level relative to the band structure of (a) a semiconductor or insulator and (b) a metal.

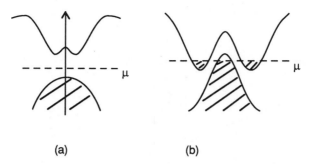

(a)                                    (b)

**FIGURE 4.24.**    Positioning of the Fermi level relative to the band structure of (a) an indirect gap semiconductor and (b) a semi-metal.

cannot be made to lase. Semiconductors that can lase must always be direct gap. That is, the energy minimum between the conduction and valence bands must occur at the momentum $k = 0$. For such materials, one can always expand the energy dependence about the minima to first order as a quadratic in $k$, the momentum.

We now wish to consider optical interactions in semiconductor materials. Consider the three processes illustrated in Figure 4.25. We wish to consider the probabilities of occurrence of each of these processes of spontaneous emission, stimulated absorption, and stimulated emission as functions of the Fermi distribution functions. Each of the processes will link a valence state of energy $\epsilon_v$ with a conduction state of energy $\epsilon_c$, where one can relate $\epsilon_v$ to $\epsilon_c$ by

$$\epsilon_c - \epsilon_v = \epsilon_g + \frac{\hbar^2 k^2}{2m_c} + \frac{\hbar^2 k^2}{2m_v} = \hbar\omega \tag{4-52}$$

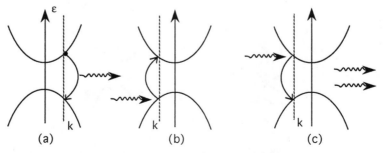

(a)                          (b)                          (c)

**FIGURE 4.25.**    (a) Spontaneous emission, (b) stimulated absorption, and (c) stimulated emission.

where $\hbar\omega$ is the associated photon energy, $m_c$ is the effective mass of the conduction electron, and $m_v$ the effective mass of the valence electron. As photons do not carry momentum, the $\epsilon_c$ and $\epsilon_v$ states must have the same $k$ values, and therefore the processes need only be parametrized by a single $k$. For spontaneous emission, it is clear that one would need an electron in the conduction band and an opening in the valence band at the same $k$ value for the process to occur. This proportionality could be expressed in the form

$$P\left(\begin{array}{c}\text{spont}\\\text{em}\end{array}\right) \sim f(\epsilon_c)[1 - f(\epsilon_v)] \tag{4-53}$$

Clearly, the probabilities of the stimulated processes must be similar to (4-53) but must include the photon number which serves as a pump. These proportionalities can be stated as

$$P\left(\begin{array}{c}\text{stim}\\\text{abs}\end{array}\right) \sim f(\epsilon_v)[1 - f(\epsilon_c)]n_p \tag{a}$$

$$P\left(\begin{array}{c}\text{stim}\\\text{em}\end{array}\right) \sim f(\epsilon_c)[1 - f(\epsilon_v)]n_p \tag{b} \quad (4\text{-}54)$$

where $n_p$ is the photon number $b^*b$.

At this point, it is possible to start making identifications of parameters in Equations (4-1) with the associated semiconductor parameters. The $\mu$ in Equation (4-1) is a parameter which refers to which atom the excited electron belongs. Clearly, from the above optical interaction discussion, the corresponding semiconductor quantity must be the $k$ of the electron. As with the atomic case, where a photon could only excite an electron in a given atom (that is, it could not replace an electron into another atom), a photon can only excite an electron in a given $k$ state. To get the constants $\gamma$ and $\tau_d$ from Equation (4-1), consider the compound process depicted in Figure 4.26. The stimulated absorption of (a) of the figure is a process that, as in the atom, causes a dipole moment to arise in this $k$ state. This dipole moment could relax either by the electron returning to its valence state or by a dephasing of the two states due to collisions. As it turns out here, the second process predominates. This process is illustrated in Figure 4.26(b). The electron in the high $k$ state will rapidly be thermalized by colliding with other conduction electrons in lower $k$ states. This thermalization occurs in much less than a picosecond, yielding a value for $\tau_\alpha$ of

$$\tau_\alpha \equiv \frac{1}{\gamma} = 100 \text{ fsec} \tag{4-55}$$

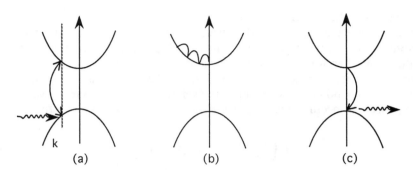

**FIGURE 4.26.**    A compound event in which (a) a photon is absorbed in a $k$ state, (b) the newly excited conduction electron scatters with the other conducting electrons to come to equilibrium near $k = 0$, and (c) the electron spontaneously emits to reenter the valence band in a low $k$ state.

The lifetime of the carrier, $\tau_d$, is much longer. As a matter of fact, the process depicted in Figure 4.26(c) takes on the order of a nanosecond, or four orders of magnitude longer than the dephasing time $\tau_\alpha$. This can be expressed as

$$\tau_d \cong 1 \text{ nsec} \tag{4-56}$$

The other identification we can make at this point is that of $d_0$. The atomic equilibrium $d_\mu = d_0$ essentially gave the probability that a photon which interacted with an atom $\mu$ would be absorbed or amplified by the $\mu^{\text{th}}$ atom. If $d_\mu$ were $-1$, it would be absorbed; if $d_\mu$ were 1, it would be amplified. But we have exactly such a quantity, with equivalent normalization, in terms of our Fermi functions. One could therefore identify $d_0$ with the difference of the $n_p$ coefficients in (4-53a) and (4-53b) to find

$$d_0 = f(\epsilon_c)[1 - f(\epsilon_v)] - f(\epsilon_v)[1 - f(\epsilon_c)]$$
$$= f(\epsilon_c) - f(\epsilon_v) \tag{4-57}$$

Unfortunately, examination of (4-57) indicates that $d_0$ must always be negative for $\epsilon_v < \epsilon_c$, and that inversion can therefore never be achieved. We realize, then, that we are still lacking an important piece of the laser dynamics, which we shall pick up in the next paragraph.

The whole point to having a semiconductor injection laser is that current can "inject" carriers into the lasing region. A typical double heterostructure is depicted in Figure 4.27. Various refinements to this structure will be discussed in the last section of this chapter. The idea behind this structure is that the alloying of GaAs with Al causes the bandgap to widen and the index

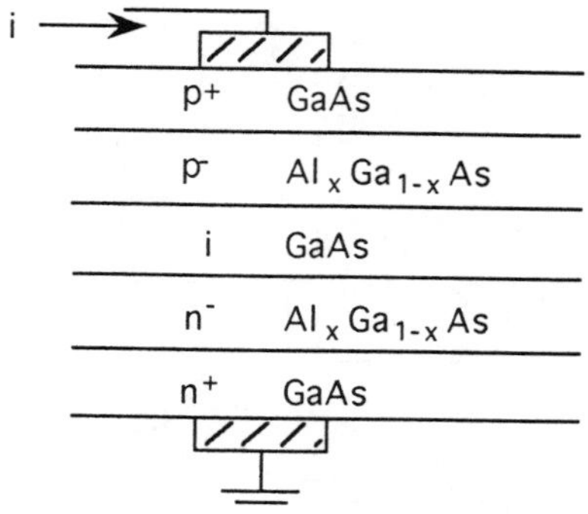

i →

| p+ | GaAs |
| p⁻ | Al$_x$Ga$_{1-x}$As |
| i | GaAs |
| n⁻ | Al$_x$Ga$_{1-x}$As |
| n+ | GaAs |

**FIGURE 4.27.**     Typical laser structure in which current is injected into the structure through electrodes bonded to $p^+$ and $n^+$ GaAs regions, which in turn inject the current through $p^-$ and $n^-$ Al$_x$Ga$_{1-x}$As regions into an intrinsic GaAs active layer.

to decrease, as illustrated in Figure 4.28. If one denotes the coordinate pointing from the lower ($n^+$) electrode to the upper electrode by $y$ and plots the gap and index as a function of this coordinate, one would obtain plots as are sketched in Figure 4.29. The basic idea here is that the GaAs intrinsic region looks not only like a waveguide layer but also like a carrier trapping region. Electrons injected from the $n^-$ region feel a barrier to enter the $p^-$ region just as do holes from the opposite direction. If one applies a voltage to the electrodes which exceeds the $\epsilon_g/e$, where $e$ is the electron charge, then current will flow into the junction. For a current $i(t)$, $i(t)/e$ carriers per unit time will flow into the junction. The potential barriers at either end of the $i$ region will cause recombination to become a dominant current mechanism. Each one of the recombination events in a sufficiently pure material will

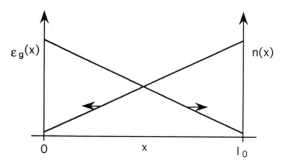

**FIGURE 4.28.**     Variation of the gap energy and index of refraction for bulk Al$_x$Ga$_{1-x}$As samples as a function of the aluminum fraction $x$.

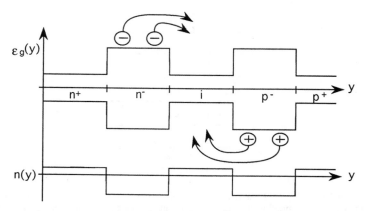

**FIGURE 4.29.** Valence and conduction band edges and the index distribution as a function of the depth coordinate $y$.

generate a photon by spontaneous emission. Photons which propagate along the $z$ direction (along the junction toward the earlier discussed cleaved end face) will tend to be bound by the index discontinuity and have a probability of being reflected back into the active region and becoming available for stimulated processes. Of course, the probability of such processes would be increased were the photons to be guided in two dimensions. This is, indeed, the subject of the final section of this chapter. The more important point here is that the injection leads to a situation in which there are free electrons, holes, and photons in the junction region that can interact. If we could identify all the parameters in the semiclassical laser equations, then we would have some handle on the dynamics of this highly nonequilibrium situation.

In the last paragraph, we tentatively identified the $d_0$ with the difference in the Fermi distributions. In the case of injected carriers, however, the Fermi distributions must be qualitatively changed. Clearly, if the volume of the intrinsic region is $V$, then the density of carriers in the conduction band $N_c$ is given by

$$N_c = \frac{I\tau_d}{eV} \tag{4-58}$$

The density of carriers in a noninjected semiconductor must be given, by definition, by

$$N_c(I=0) = \int_0^\infty \rho_c(k)f(k)\,dk \tag{4-59}$$

where $\rho_c(E)$ is the density of states in the conduction band (which we will not discuss here) and $E$ is the energy measured from the conduction band edge. For low temperatures ($\epsilon_g \gg k_B T$, which is generally true), $N_c(I = 0) \sim 0$. For a given $I$, however, $N_c$ can become large. One way to view this is as a change in the Fermi level from the value $\mu$ to a new value which we call the quasi-Fermi level $\mu_c$. The situation is somewhat as depicted in Figure 4.30, where electrons and holes are poured into their respective bands in sufficient numbers that a quasi-equilibrium is set up. In such a case, one can think of quasi-equilibrium $d_0$, which would be given by

$$d_0 = f_c(\epsilon_c) - f_v(\epsilon_v) \tag{4-60}$$

where the thermal $d_0$ term has been ignored and the $f_c(\epsilon_c)$ and $f_v(\epsilon_v)$ are given by

$$f_c(\epsilon_c) = \frac{1}{e^{(\epsilon_c - \mu_c)/k_B T} + 1} \tag{a}$$

$$f_v(\epsilon_v) = \frac{1}{e^{(\epsilon_v - \mu_v)/k_B T} + 1} \tag{b} \tag{4-61}$$

When the number of photons in the cavity is small, then the second term on the right-hand side of Equation (4-1c) is small and the semiconductor version of this equation will admit solutions where $d_k = d_0$. For *positive* values of $d_0$, however, the photon number will be large and $d_k$ will be determined from system dynamics. One could, however, still use the equations

$$N_c = \int_0^{-\infty} \rho_c(k) f_c(k) \, dk \tag{a}$$

$$N_v = \int_0^{\infty} \rho_v(k)(1 - f_v(k)) \, dk \tag{b} \tag{4-62}$$

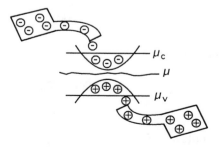

**FIGURE 4.30.**    Injection process in a semiconductor.

where $\rho_c(k)$ and $\rho_v(k)$ are the densities of $k$ states in the conduction and valence bands respectively,

$$N_c = \frac{I_e \tau_d}{eV} \qquad \text{(a)}$$

$$N_v = \frac{I_h \tau_d}{eV} \qquad \text{(b)} \quad (4\text{-}63)$$

where the total current $I$ equals the electron current $I_e$ which in turn equals the hole current $I_h$, to determine the conduction band quasi-Fermi level $\mu_c$ and the valence band quasi Fermi level $\mu_v$. With these, one could use Equation (4-60) to define the "quasi-equilibrium" $d_0$. This $d_0$ could then be used to drive our present system of equations

$$\frac{db_\lambda}{dt} = (-i\omega_\lambda - \kappa_\lambda)b_\lambda - i \sum_k g^*_{k\lambda} \alpha_k + F_\lambda(t) \qquad \text{(a)}$$

$$\frac{d\alpha_k}{dt} = \left(-i\omega_k - \frac{1}{\tau_\alpha}\right)\alpha_k + i \sum_\lambda g_{k\lambda} b_\lambda d_k \qquad \text{(b)}$$

$$\frac{dd_k}{dt} = \frac{d_0 - d_k}{\tau_d} + 2i \sum_\lambda (g^*_{k\lambda} b^*_\lambda \alpha_k - g_{k\lambda} b_\lambda \alpha^*_k) \qquad \text{(c)} \quad (4\text{-}64)$$

where all the parameters, with the exception of the $g$'s, have been defined in the preceding discussion.

An interesting question concerns the threshold current density necessary to achieve inversion. As the current density determines the quasi-Fermi levels, it is easier to discuss the conditions on the quasi-Fermi levels. For inversion to be achieved requires that the $d_0$ be greater than zero or, from (4-60) and (4-61), that

$$\frac{1}{e^{(\epsilon_c - \mu_c)/k_B T}} > \frac{1}{e^{(\epsilon_v - \mu_v)/k_B T}} \qquad (4\text{-}65)$$

where the $\epsilon_c$ and the $\epsilon_v$ must be related by

$$\epsilon_c - \epsilon_v = \epsilon_g + \frac{\hbar^2 k^2}{2m_c} + \frac{\hbar^2 k^2}{2m_v} \qquad (4\text{-}66)$$

for a state with momentum $k$. Solving (4-65) for $\mu_c - \mu_v$, one finds that for a state of momentum $k$ to be inverted requires that

$$\mu_c - \mu_v > \epsilon_g + \frac{\hbar^2 k^2}{2m_c} + \frac{\hbar^2 k^2}{2m_v} \tag{4-67}$$

or that, for a state at $k = 0$ to be inverted, the Fermi levels must be separated by more than the gap energy. This relation can be used to explain why the indirect gap semiconductor cannot be made to lase. Figure 4.31 is an attempt to illustrate the explanation. In the direct gap semiconductor, as soon as the pump current becomes high enough to push $\mu_c$ above the conduction gap edge and $\mu_v$ below the valence band edge, these "inverted" states are free to take part in purely optical interactions. The situation is clearly quite different in the indirect gap semiconductor. Here one could push both the Fermi levels into their respective bands and still have no purely optical interactions taking place, as the $k$ values of the "free" states would not match. The Fermi levels would have to be pushed much deeper into their respective bands to achieve purely optical interactions. However, before this could occur, other effects could take place. For example, the occupied conduction band states could be connected with the $k = 0$ valence band states by the simultaneous emission of a photon and a phonon of momentum $k$. Although this higher order interaction is of much lower probability than the first order photon interaction, it will occur as it is the lowest order game in town. This interaction is rather disastrous to the material, as phonons are heat. As more phonons are generated in the lattice, the interaction probability goes up and more phonons are generated, causing thermal runaway and the destruction of the crystal. Indeed, no one has ever been able to make an indirect gap semiconductor (e.g., silicon) lase. (Through doping, lasing can occur in a host which has an indirect bandgap, but it is the dopants that are brought to threshold.)

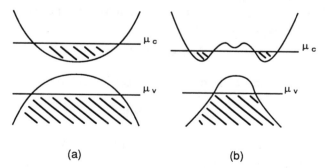

(a)                                    (b)

**FIGURE 4.31.**    (a) A critically pumped direct bandgap semiconductor and (b) a critically pumped indirect bandgap semiconductor.

## 4.4 SEMICONDUCTOR RATE EQUATIONS

We are now ready to put Equations (4-64) into their most commonly used form, that of the semiconductor rate equations. We start this derivation with the observation that the photon lifetime $\tau_p$ is greater than the dephasing time $\tau_\alpha$. As the $\alpha_k$ equation of (4-64b) is a driven equation, a state which is excited by a photon of $\omega_\lambda$ will initially oscillate at $\omega_\lambda$ even if the transition corresponds to a slightly different frequency $\omega_k$. This discrepancy is not only allowed but is a basic one to quantum theory, which explains linewidths. As $\tau_\alpha$ is small, the polarization state will decay before it has time to select its own oscillation frequency. We can therefore assume that

$$\frac{d\alpha_k}{dt} = -i\omega_\lambda \alpha_k(t) \tag{4-68}$$

to obtain the relation

$$\alpha_k(t) = \frac{i \sum_\lambda g_{k\lambda} b_\lambda d_k}{i(\omega_k - \omega_\lambda) + \gamma} \tag{4-69}$$

Using (4-69) in (4-64a) and (4-64c) gives one the coupled system

$$\frac{db_\lambda}{dt} = (-i\omega_\lambda - \kappa_\lambda)b_\lambda + \sum_{k,\lambda'} \frac{g_{k\lambda}^* g_{k\lambda'} b_{\lambda'} d_k}{i(\omega_k - \omega_\lambda) + \gamma} + F_\lambda(t) \tag{a}$$

$$\frac{dd_k}{dt} = \frac{d_0 - d_k}{\tau_d} - 4\left[\sum_{\lambda'} \frac{\gamma g_{k\lambda}^* b_\lambda^* g_{k\lambda'} b_{\lambda'} d_k}{(\omega_k - \omega_\lambda)^2 + \gamma^2}\right] \tag{b} \quad (4\text{-}70)$$

Equation (4-70) is close to the form we desire but still slightly different. Before putting in the rate equation form, however, we wish to investigate what the desired form is. In classical electromagnetism, one usually defines a polarization $\mathbf{P}$ in terms of

$$\mathbf{D} = \epsilon_0 \mathbf{E} + \mathbf{P} \tag{4-71}$$

If one assumes that this $\mathbf{P}$ is linear in the electric field, one can obtain

$$\mathbf{D} = \epsilon_0(1 + \chi)\mathbf{E} \tag{4-72}$$

If one assumes that the electromagnetic disturbances are made up of transverse plane waves of the form

$$\mathbf{E} = \mathbf{E}_t e^{i\frac{\omega}{c}z} \tag{4-73}$$

then the assumption of (4-72) leads to a wave equation of the form

$$\frac{\partial^2 E_t}{\partial t^2} = -\frac{\omega^2}{(1+\chi)} E_t \tag{4-74}$$

If one were to restrict consideration to only positive frequency waves, one could obtain an equation for $b(t)$, the positive frequency component coefficient of the wave, of the form

$$\frac{\partial b_\lambda}{\partial t} = -\frac{i\omega_\lambda}{\sqrt{1+\chi}} b_\lambda \tag{4-75}$$

Expanding equation (4-75) for small $\chi$, one obtains

$$\frac{\partial b_\lambda}{\partial t} = \left(-i\omega_\lambda + \frac{i\omega\chi}{2}\right) b_\lambda \tag{4-76}$$

In general, $\chi$ could be complex. Further, there can be cavity losses given by $\kappa_\lambda$. Plugging these factors into (4-74), one finds the suggestive form

$$\frac{\partial b_\lambda}{\partial t} = (-i\omega_\lambda - \kappa_\lambda)b_\lambda + \frac{i\omega\chi}{2} b_\lambda \tag{4-77}$$

This form includes the static effects of the resonant cavity frequency and cavity loss, as well as the dynamical effects of frequency pulling ($\chi_r$) and dynamical absorption or gain ($\chi_i$). For a classical spring model of matter, the $\chi$ can be determined from the Classius-Mosetti relations.

The question remains as to what we must have to do to equations (4-70a) and (4-70b) to obtain (4-70a) in a form such as that of (4-77). Clearly, by considering single mode operation, one can set $\lambda = \lambda'$ in (4-68) to obtain

$$\frac{\partial b_\lambda}{\partial t} = (-i\omega_\lambda - \kappa_\lambda)b_\lambda + \sum_k \frac{|g_{k\lambda}|^2 d_k}{i(\omega_k - \omega_\lambda) + \gamma} b_\lambda + F_\lambda(t) \quad \text{(a)}$$

$$\frac{\partial d_k}{\partial t} = \frac{d_0 - d_k}{\tau_d} + 4\frac{\gamma|g_{k\lambda}|^2 d_k}{(\omega_k - \omega_\lambda)^2 + \gamma^2} n_\lambda \qquad \text{(b)} \quad \text{(4-78)}$$

One more simplification can be used to "improve" the form of (4-78). This requires letting the sums over $k$ go over to weighted integrals according to

$$\sum_k \rightarrow \int d_k \rho(k) \tag{4-79}$$

where $\rho(k)$ is the earlier mentioned density of states in $k$ space. This substitution can be performed on the sum term of Equation (4-78a) immediately. But we also wish to obtain a "summed" form of (4-78b) in which the variable will be

$$D(t) = \int \rho(k) d_k \, dk \tag{4-80}$$

The only unknown in (4-78b) would then be the $k$ sum over $d_0$. This sum can be obtained as follows:

$$\int \rho(k) d_0 \, dk = \int \rho(k) [ f_c(\epsilon_c) - f_v(\epsilon_v) ] \, dk \tag{4-81}$$

by the definition of $d_0$. However, one knows that $N_v$, the density of holes, must be equal to $N_c$, the density of conduction electrons, which in turn must be given by

$$N_c = N_v = \frac{I \tau_d}{eV} \tag{4-82}$$

where the right-hand side of (4-82) gives the number of injected carriers. The $N_c$ and $N_v$, however, can also be defined from the relations

$$N_c = \int \rho(k) f_c(\epsilon_c) \, dk \tag{a}$$

$$N_v = \int \rho(k) [ 1 - f_v ] \, dk \tag{b} \tag{4-83}$$

Combining (4-81) and (4-83), one finds that

$$\int \rho(k) d_0 \, dk = N_c + N_v - N \tag{4-84}$$

where $N$ is the total number of states in a band defined by

$$N = \int \rho(k) \, dk \tag{4-85}$$

Using (4-80), (4-84), and (4-82) in (4-78b) and (4-79) in (4-78a), one obtains the system

$$\frac{db_\lambda}{dt} = (-i\omega_\lambda - \kappa_\lambda) + i\tilde{\chi}b_\lambda + F_\lambda(t) \tag{a}$$

$$\frac{dD}{dt} = \frac{1}{\tau_d}\left(\frac{I\tau_d}{2eV} - N - D(t)\right) - 2\chi_i n_\lambda \tag{b} \quad (4\text{-}86)$$

where $\tilde{\chi}$ is defined by

$$\tilde{\chi} = i\int dk\,\frac{|g_{k\lambda}|^2 d_k}{i(\omega_k - \omega_\lambda) + \gamma}\,\rho(k)$$

$$= \chi_r + i\chi_i$$

$$= 2\int dk\,\frac{(\omega_k - \omega_\lambda)|g_{k\lambda}|^2 d_k}{(\omega_k - \omega_\lambda)^2 + \gamma^2}\,\rho(k) - 2i\int dk\,\frac{\gamma|g_{k\lambda}|^2 d_k}{(\omega_k - \omega_\lambda)^2 + \gamma^2}\,\rho(k) \quad (4\text{-}87)$$

It should be noted here that $N + D(t)$ is often denoted as $N(t)$ in the literature and Equation (4-86b) then written in terms of this quantity instead of $D(t)$ and $N$. Note that a noninverted medium would have $d_k < 0$ and that therefore the noninverted $\chi_i$ is positive and therefore damps the field while serving to pump the inversion.

The important purpose of this section is to compare the $\chi$'s for the atomic model and the semiconductor model and therefore exhibit the salient differences between gas laser and semiconductor laser operation. As could be shown from the discussion in the last chapter (e.g., Equation 3-100), the real and imaginary parts of $\chi$ for a homogeneously broadened atomic medium laser take the form of a "pure" Lorentzian

$$\chi = +\frac{ig^2(N_b - N_a)}{i(\omega - \omega_\lambda) + \gamma} \tag{4-88}$$

as is sketched in Figure 4.32 for $N_b < N_a$, the "usual" condition, that is, for a noninverted medium. Perhaps the most important point about this is that the absorption maximum occurs at the point that the $\chi_r$ has a zero. This is to say that the absorption peaks around a point where there is no dynamical index contribution to the background index. As the only change to these curves of Figure 4.32 that occurs with inversion is a sign change, a similar situation will occur under lasing conditions. As a laser wants to lase at the point of maximum gain, one could say that an atomic laser will lase at or near

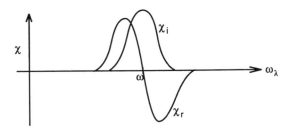

**FIGURE 4.32.**    "Standard" atomic polarizability curve around a resonance.

the point where the dynamical contribution to the background index is zero. This is an important characteristic of a gas laser for the following reason. As has been discussed, the empty cavity conditions determine the allowed frequencies of the cavity modes. Any change of index in the cavity will tend to steer these frequencies and thereby generate phase noise. As the gain peak of the atomic medium always sits on top of the zero of the dynamic index, however, changes in gain will not induce additional phase noise. As we shall soon see, this is not the case in the semiconductor laser.

As we have not yet rigorously defined either the $g_{k\lambda}$ nor the $\rho(k)$ and do not intend to do so, this discussion will be somewhat qualitative. The reader can be referred to the literature for more information (see, for example, Yariv 1989, Chapter 15). Perhaps a first point to make is that $\chi$ for emission will not take on the same form apart from a minus sign as $\chi$ for absorption, because $d_0$, which is given by

$$d_0 = \frac{1}{e^{(\epsilon_c - \mu_0)/k_b T} + 1} - \frac{1}{e^{(\epsilon_v - \mu_v)/k_b T} + 1} \tag{4-89}$$

is clearly not antisymmetric in the quantity $\mu_c - \mu_v - \epsilon_g$. The absorption spectra will therefore be considerably different from the gain spectra under inversion. Further, the operating point will not be fixed. This can be seen from perusal of the expression for $\chi_i$ in Equation (4-87). As the damping $\gamma$ must be small, the Lorentzian line shape function must be sharply peaked around $\omega_k = \omega_\lambda$ and therefore one can approximate

$$\chi_i \sim -d(\omega_\lambda)\rho(\omega_\lambda) \tag{4-90}$$

where constants have been ignored. This function is sketched for three values of $\mu_c - \mu_v$ in Figure 4.33. The important features of the plot are determined by the fact that $\chi_i(\omega_g) = 0$ from $\rho(\omega_\lambda < \omega_g) = 0$. That is, there are no states in

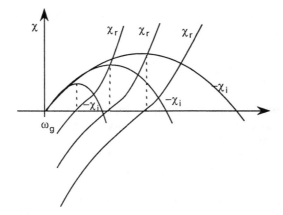

**FIGURE 4.33.**    Variations of $\chi_r$ and $\chi_i$ with the frequency $\omega_\lambda$.

the gap. Further, we know that $\chi_i(k_\mu) = 0$ where $k_\mu$ is defined by

$$\mu_c - \mu_v - \epsilon_g + \frac{\hbar^2 k_\mu^2}{2m_c} + \frac{\hbar^2 k_\mu^2}{2m_v} = 0 \tag{4-91}$$

In between these values, the shape of the gain curve is probably somewhat parabolic. The integral for the real part of $\chi$ is more nasty to do, but one can assume that the complex function $\chi$ is analytic to validate the Kramers-Kronig relation (see, for example, Yariv 1989, Appendix 1). These relations simply state that $\chi_r$ and $\chi_i$ must form a Hilbert transform pair. Three sketches of the Hilbert transforms of the $\chi_i$'s are included in Figure 4.33 and labeled as $\chi_r$'s. As in the atomic case, the zeroes of the $\chi_r$'s coincide with the gain maxima. However, these zero frequency points are now a function of pumping. As the pump increases, the gain maxima shifts to a higher frequency, therefore dragging the photon frequency with it. The cavity phase condition must be satisfied, however, for a laser mode. As the gain peak moves, the modal frequency cannot track it precisely. For example, if the cavity frequency coincided with the peak of the smallest $\chi_i$ curve in Figure 4.33, an increase in pump would shift this peak to first order through the change in $\chi_r$ at the maxima, which would decrease from zero, rather than due to the change in $\chi_i$. The effect of the gain change would be more likely to cause mode hopping than chirp. However, the point here is that the gain and phase noise are strongly coupled. This fact causes both the commonly observed turn-on chirp of semiconductor lasers as well as an enhanced linewidth of the semiconductor laser relative to the Schalow-Townes formula (see, for example, Yariv 1989, Chapter 21).

## 4.5 SEMICONDUCTOR LASER STRUCTURES

A practical laser structure was given in Figure 4.27. Here the purpose is to discuss in some detail how such a structure can be made and to apply the effective index method to optimization of the structure.

The typical dimensional parameters of a laser are given in Figure 4.34. The first thing evident from the drawing is that it is not to scale. The point here is really that the 300 $\mu$m length is actually an important parameter, as it determines the cavity length. The two dimensions transverse to this are arbitrary and are generally picked only for mechanical stability of the structure. The important transverse dimensions are those of the channel, and they are considerably smaller than the channel length. The roughly 0.2 $\mu$m channel depth is essentially determined by the requirement that, the thinner the intrinsic recombination region, the higher the recombination rate and therefore the lower the threshold current. The other transverse dimension of the junction should ideally be kept as close to this depth value as possible but, as we shall soon see, this is something of a technological problem.

With reference to Figure 4.27, some discussion should be given to the growth of the layers. The $n^+$ GaAs layer is quite generally the substrate layer. This layer is usually bought from a mass producer of GaAs boules (macroscopic synthetically grown crystals) and is of much lower quality than the epitaxially grown layers above it. The point is that, as long as this layer is heavily doped, it simply acts as an extension of the lower contact. As no light will be guided in it, it need not be of ultrahigh purity. It can also be thick ($\geq 0.5$ mm) for mechanical stability, and as cheap as possible. The $p^-$ and $n^-$ Al$_x$Ga$_{1-x}$As layers need to be of high quality and thick enough ($\geq 1$ $\mu$m) to completely buffer the high field intrinsic GaAs layer from the high optical

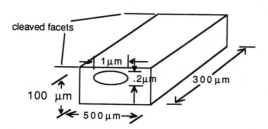

**FIGURE 4.34.**    Semiconductor laser structure.

loss $p^+$ and $n^+$ contact layers. As was mentioned previously, the intrinsic layers need to be thin ($\leq 0.2$ $\mu$m) so that the injected carriers may recombine as efficiently as possible. The lowest threshold density lasers are ones in which this layer is actually a simple quantum well of hundreds of angstroms in depth in which the electron wavefunctions are actually distorted by the effective potential well.

As was mentioned previously, one buys the substrate but must grow the laser structure on this substrate, with the additional requirement that materials that are grown must be of very high purity. The science of doing this (which many consider more an art than a science) is that of epitaxy or epitaxial growth. There exist various techniques to carry it out, of which we will mention three. The oldest of these, and the one still used for most commercial lasers, is *liquid phase epitaxy* (LPE). In this method, one simply immerses the sample in a bath which is of a proper composition to crystallize the correct material on the sample's surface. This technique is easily implemented but is the dirtiest of those mentioned here and the one which allows least control of the device geometry. A second technique is *molecular beam epitaxy* (MBE). In this method, the sample is placed in a vacuum chamber, and molecular beams of the elements of the desired composite are trained on the surface. The third method is *metal-organic chemical vapor deposition* (MOCVD). In this technique, gaseous organic compounds containing the desired elements of the composition are reacted on top of the surface on which one wishes to perform the deposition. Chemical reactions take place, leaving the correct layer deposited on the surface and some nasty gaseous byproducts that need to be exhausted from the reactor vessel. Accurate, tunable reaction rates can be achieved with this technique. Unfortunately, a number of the gases used and produced by MOCVD are extremely toxic, which limits the areas of applicability of the technique.

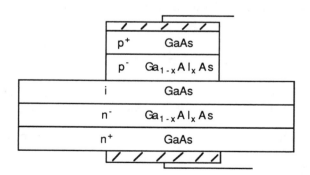

**FIGURE 4.35.**    Layer structure of a mesa etched semiconductor laser structure.

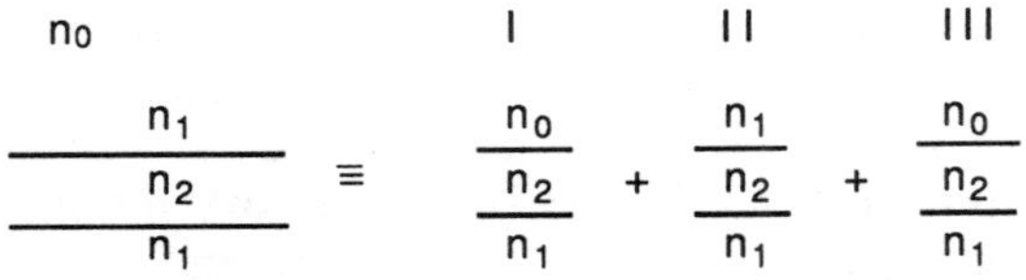

$$\begin{array}{c} n_0 \\ \hline n_1 \\ \hline n_2 \\ \hline n_1 \end{array} \equiv \begin{array}{c} \mathbf{I} \\ n_0 \\ \hline n_2 \\ \hline n_1 \end{array} + \begin{array}{c} \mathbf{II} \\ n_1 \\ \hline n_2 \\ \hline n_1 \end{array} + \begin{array}{c} \mathbf{III} \\ n_0 \\ \hline n_2 \\ \hline n_1 \end{array}$$

**FIGURE 4.36.**    The laser structure of Figure 4.35 broken into its effective index segmentation.

Any of these techniques can be used to grow a structure like the one depicted in Figure 4.27. There is, however, a double problem with this structure. Although both carriers and fields are confined in the depth dimension, there is no transverse confinement of either. Here we wish to discuss one method of obtaining some transverse confinement, called *mesa etching*. In mesa etching, after a structure like that of 4.27 is grown, one can mask it and apply either wet chemical etch, ion milling, or reactive ion etching and then remove the resist to obtain a structure like the one in Figure 4.35. It is clear that this structure will have a controlling effect over injected current, as it certainly restricts the current paths through the $p^-$ and $p^+$ regions. It is not so obvious that it also leads to transverse confinement. The following effective index argument should clarify the matter. For purposes of the exposition, we do not include the $p^+$ and $n^+$ regions, as, if the fields reach these regions and are therefore strongly attenuated, we have found a poor solution anyway and need to try again. We now split our guidance problem into four problems, the first three of which are described in Figure 4.36. The beta values that will be obtained for these slab problems relative to $n_0 k_0$, $n_1 k_0$, and $n_2 k_0$ are depicted in Figure 4.37. The most important point of this exercise is that the fourth problem of the effective index decomposition of the guide structure of Figure 4.35, the transverse guidance problem, becomes a symmetric slab structure, as illustrated in Figure 4.38. A symmetric slab structure always admits at least one bound mode. Therefore, the effective

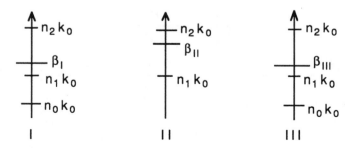

**FIGURE 4.37.**    Positions of the $\beta$'s of the slab problems of Figure 4.36.

$$n_I = \frac{\beta_I}{k_0} \qquad\qquad n_{II} = \frac{\beta_{II}}{k_0} \qquad\qquad n_{III} = n_I = \frac{\beta_I}{k_0}$$

**FIGURE 4.38.**    The fourth slab problem of the effective index decomposition of the structure of Figure 4.35, that of the transverse guidance of the structure.

"loading" of the beta values of problems I and III due to the etching has caused the structure to become transversely guiding. The magnitude of the decay in regions I and III, given by

$$K_I = K_{III} = \sqrt{\beta^2 - n_1^2 k_0^2} \tag{4-92}$$

could be small, indicating only weak confinement. This is, in fact, a problem with mesa structures, but one whose solution is the subject of other books. The basics presented here should be enough to allow one to delve into the laser literature.

## References

Abramowitz, M. and Stegun, I., 1965. *Handbook of Mathematical Functions.* Dover Publications, Inc., New York.

Brillouin, L., 1960. *Wave Propagation and Group Velocity.* Academic Press, New York.

Goell, J.E., 1969. "A Circular-Harmonic Computer Analysis of Rectangular Dielectric Waveguides." *Bell Sys. Tech. J.,* **48,** pp. 2133–2160.

Knox, R.M. and Toulios, P.P., 1970. "Integrated Circuits for the Millimeter Through the Optical Frequency Range." *Proceedings of the Symposium on Submillimeter Waves, Polytechnic Institute of Brooklyn* (March 31–April 2, 1970), p. 497.

Kittel, C., 1971. *Introduction to Solid State Physics,* Fourth Edition. John Wiley & Sons, New York.

Maliuzhinetz, G.D., 1958. "Excitation, Reflection, and Emission of Surface Waves from a Wedge with Given Face Impedances." *Dokl. Akad. Nauk SSSR,* **121,** pp. 436–439.

Marcatilli, E.A.J., 1969. "Dielectric Rectangular Waveguide and Directional Coupler for Integrated Optics." *Bell Sys. Tech. J.,* **48,** pp. 2079–2102.

Watson, G.N., 1966. *A Treatise on the Theory of Bessel Functions.* Cambridge University Press, Cambridge.

Yariv, A., 1989. *Quantum Electronics,* Third Edition. Holt, Rinehart & Winston, New York.

## PROBLEMS

1. Consider the structure depicted in Figure 4.39, where one can take $n_1 > n_2 > n_0$ and $\Delta n \geq 0$. Consider each region to be square and of side $a$.

   (a) Write down the form of the equations which would be used to define the effective index solution for this structure.

   (b) For $\Delta n = 0$, sketch what the fundamental mode of the structure would look like for $(n_1 - n_2)a \approx \lambda$, and $(n_1 - n_2)a \ll \lambda$, but $(n_1 - n_0)a \geq \lambda$.

   (c) Taking $(n_2 = n_1)a \geq \lambda$, sketch how the fundamental mode would vary as $\Delta n$ varies from 0 to $2(n_1 - n_2)$.

2. Consider the waveguide depicted in Figure 4.40. Write out what the form of the effective index solution for what the fundamental mode would be in each of the regions 1–9. Sketch the form of these solutions along the lines $A$ (at $0°$) and $B$ (at $45°$).

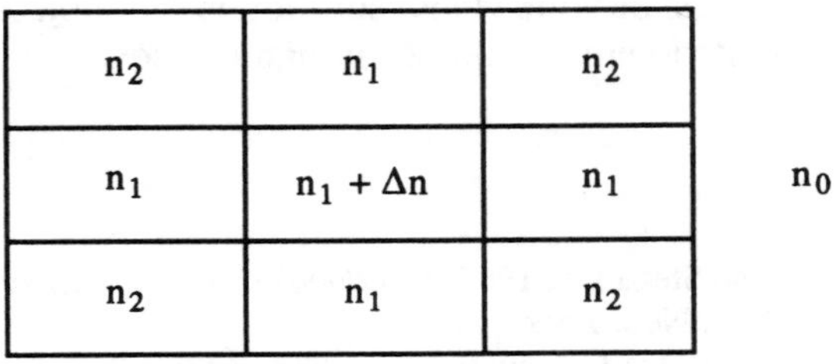

| $n_2$ | $n_1$ | $n_2$ | |
|-------|-------|-------|-------|
| $n_1$ | $n_1 + \Delta n$ | $n_1$ | $n_0$ |
| $n_2$ | $n_1$ | $n_2$ | |

**FIGURE 4.39.**

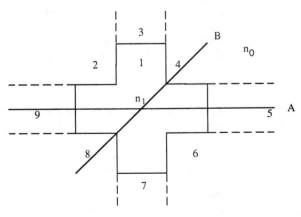

**FIGURE 4.40.**

3. In the text it was shown that for a piecewise continuous medium the transverse fields $E_x$, $H_x$, $E_y$, and $H_y$ could all be expressed as derivatives of the normal fields $E_z$ and $H_z$ where $E_z$ and $H_z$ satisfied

$$\frac{d^2E_z}{dx^2} + \frac{d^2E_z}{dy^2} + (k_i^2 - \beta^2)E_z = 0$$

$$\frac{d^2H_z}{dx^2} + \frac{dH_z}{dy^2} + (k_i^2 - \beta^2)H_z = 0$$

It was mentioned in the text that for a general 2-dimensional index $n(x,y)$, it is possible to express the transverse fields $E_x$, $E_y$, $H_x$, $H_y$ in terms of the longitudinal fields $E_z$ and $H_z$ and then find coupled equations for $E_z$ and $H_z$. Find the transverse-longitudinal relations and coupled equations for the longitudinal fields. Try to find index profiles for which you can solve these equations, or at least make statements about the field structure.

4. Consider the transverse index distribution sketched in Figure 4.41. Assume that $n_1 > n_2$.
   (a) Try to find the form of the modal solutions in each of the regions.
   (b) Write down the boundary conditions for a modal solution.
   (c) Between what values of $k$ will the eigenvalue (propagation constant) $\beta$ be bound?
   (d) Sketch what the first few modes will look like.
5. Although we generally think of applying the effective index method to

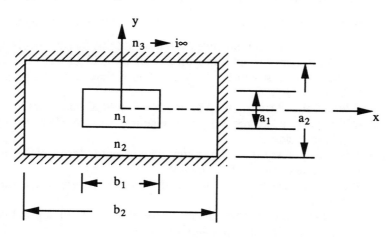

**FIGURE 4.41.**

rectangular-type geometries, one could apply it to the circular dielectric rod. A great advantage of the method is that one can sketch the shapes of the eigenfunctions to the various problems without calculating the $\beta$'s in any horrible detail. Make a gross discretization of a dielectric cylinder problem and sketch the rough form of the wave functions for the associated problems. When finished, having thereby obtained the "true" 2-dimensional wave function, sketch this function on various 1-dimensional slices cut through the origin at various angles from 0 to $\pi/2$. Do the wave functions have circular symmetry?

6. In Chapter 3, we used symmetry properties of spatial wave functions to generate equations coupling $C_a(t)$ and $C_b(t)$. We also know the shape of the $u_k$ and $u_{-k}$ wave functions (above and below the bandgap) from Chapter 4. Go step by step through the arguments in Chapter 3 but using the $C_k$ and $C_{-k}$ coefficients to find equations for $\alpha_k$ and $d_k$. You should find some severe mathematical problems along the way which you should point out.

7. Solve the following:
   (a) Say we are given that the static dielectric constant of GaAs is 13, the electrical force between an electron and a hole is given by $\dfrac{e^2}{\epsilon r^2}$, in cgs units, in a crystal lattice, that the effective electron mass in GaAs is $0.05m_e$ and that the hole mass in GaAs is much greater than the electron mass. Use the uncertainty relation to calculate the exciton radius in GaAs, that is, the radius of the lowest-order wavefunction of a bound electron-hole pair. Evaluate this radius in angstroms.

   (b) Say, instead of using the rest mass of the electron to calculate $\Delta p$ as in (a), we used $\epsilon_{thermal} = \dfrac{3kT}{2} \approx \dfrac{p^2}{2m}$ to calculate the thermal uncertainty in the electron position. What would this thermal de Broglie wavelength be at room temperature? At $10^2$ °K? At $10^{12}$ °K? How would you choose which wavelength to use, that calculated in (a) or (b)?

8. Consider a cube of sides $a$ with an infinite potential on the walls.
   (a) Write down an expression for the eigenmodes of the Schroedinger equation inside the cube.
   (b) Give an expression for the energy eigenvalues in terms of the principal mode number of the modes.
   (c) Find an expression for the energy degeneracy in terms of the mode number.
   (d) For large mode number, the mode numbers form an approximate continuum. Find an expression for the density of states in this limit.

What would happen to this expression were the cube to be deformed into a highly irregular shape?

9. The only difference between the Fermi-Dirac distribution (FDD) and the Bose-Einstein distribution (BED) is the use of the Pauli exclusion principle in the deviation of the FDD. In the FDD, a state is either empty or has occupancy 1, whereas in the BED, any number of particles can be present.

(a) Using the Gibbs factor

$$\frac{P(N_1,\epsilon_1)}{P(N_2,\epsilon_2)} = \frac{e^{(N_1\mu-\epsilon_1)/k_BT}}{e^{(N_2\mu-\epsilon_2)/k_BT}}$$

that is, the probability of a state with $N_1$ particles and energy $\epsilon_1$ divided by the probability of a state with $N_2$ particles and energy $\epsilon_2$ is given by a ratio of exponentials involving the particle numbers, energies, Fermi level $\mu$, Boltzmann's constant $k_B$, and temperature $T$, and the fact that the average occupancy of a state $\langle n \rangle$ is given by the sum

$$\langle n \rangle = \sum_n nP(n)$$

where $P(n)$ is the probability of the $n$ particle state, derive the FDD and the BED.

(b) Under what conditions (the classical limit) do these distributions reduce to the same expression?

(c) In our simple model of band structure, the $d_0$ was a simple function of the FDD's of the valence and conduction bands. In the classical limit ($\mu$ deep in the gap), what does $d_0$ reduce to? For states near the band edges, what does $d_0$ reduce to when $\mu_c - \mu_v > \epsilon_g$?

(d) Try to repeat (c), but for the case where the states in the valence and conduction bands are BE distributed. Does anything strange happen?

10. Recall the argument in the notes leading to the statement of the semiconductor inversion condition, $\mu_c - \mu_p > \epsilon_g$. Starting from the rate equations, fill in the missing steps in the derivation. Discuss the physical meaning of any approximations made.

11. In the text some discussion is given to how semiconductor gain curves and index change curves could be obtained. Calculate these curves in some detail for various inversion levels.

12. Say we have a wave which satisfies the equation

$$\frac{db}{dt} = i\omega(1 + \chi)b$$

propagating through a medium of susceptibility $\chi$. Sketch the frequency spectrum of this wave for:
(a) $\chi$ a constant independent of $b$
(b) $\chi(t) = \chi_0 \cos \omega_r t$
(c) $\chi(b) = \chi_0 + \chi_1 b(t)$
(d) $\chi(b) = \chi_0 + \chi_1 |b(t)|^2$

13. Classically, the susceptibility can be defined by

$$\chi = \frac{N\alpha}{1 - \dfrac{N\alpha}{3}}$$

where the $\alpha$ can be found from a spring model of the atom as

$$\alpha(\omega) = -\frac{e}{\epsilon_0} \frac{|r(\omega)|}{|E(\omega)|}$$

where $r(\omega)$ defines the motion of a mass $m$ attached to a spring of spring constant $\omega_0$ and the damping constant $\gamma m$. Now the semiconductor laser rate equation should be expressible as

$$\frac{db}{dt} = (-i\omega - k)b + \frac{i}{2} \omega\chi b$$

$$\frac{dD}{dt} = \text{constant } \chi_i |b|^2 - \frac{D_0 - D}{\tau_d}$$

By comparing the above with the Maxwell-Bloch system, one should obtain relations corresponding to the classical $\chi$ expressions. Carry out a detailed comparison of these expressions and describe the difference between the classical and the semiclassical theories.

14. Recall that the equilibrium inversion in a two-level system is given by

$$d_0 = \frac{1 - e^{\hbar\omega/kT}}{1 + e^{\hbar\omega/kT}}$$

Let us say we apply a forward bias of $V$ volts to a $p$-$n$ junction. Plot the "effective" temperature inside the junction as a function of the constant voltage. How does one interpret the curve? Can you use the relation

$$\frac{1}{k_B T} = \frac{d(\log g)}{du}$$

where $g$ is the number of available states at an energy $u$ to show that energy does not always flow from a state at higher temperature than at lower?

15. The carrier rate equation

$$\dot{N}(t) = \frac{N(t)}{\tau_d} + \frac{\omega}{n_b^2} \chi_i n_p + \frac{2I(t)\tau_d}{eV}$$

cannot be correct under transient conditions, as the pump has an instantaneous effect. Further, $I(t)$ must be a function of the recombination. This problem will try to fix up this problem by replacing the carrier equation with two dynamical equations.

(a) Try to develop an equivalent circuit for the diode junction by applying a voltage across it and calculating the resulting current as a function of time. Ignore transit time effects. Consider a circuit like that depicted in Figure 4.42.

(b) Try to fix up the pump term in the carrier equation to reflect the fact that there is a delay in the system.

(c) Try to sketch the evolution of the system after a sudden turn on of the voltage source, assuming $n_p = 0$.

(d) Qualitatively, what happens to the behavior of the system if one takes $n_p = $ constant? Is $\chi_i$ constant?

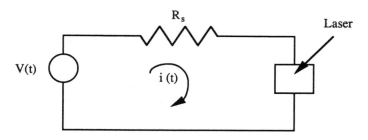

**FIGURE 4.42.**

16. Here we wish to give some consideration to the pumping problem in a semiconductor laser. As was pointed out in Chapter 4, the pumping parameter $D_0$ at a given $k$ value must be proportional to the injected current. The injected current, however, should be the solution of a circuit differential equation involving the internal resistance of the current source and the junction capacitance. Taking the junction capacitance as being independent of the pumping (a very poor approximation), write down the system of two equations, and solve it approximately for various values of the circuit $RC$ time constant. Make sketches of your results.

# 5

---

# Optical Fibers

In this chapter, the plan is to give a brief exposition of the most important aspects of propagation in optical fibers. Attention will be placed on understanding what kinds of fibers go together with what kinds of sources for what kinds of applications. The first section of the chapter contains a very brief historical overview. The second section contains solutions to both the hybrid and LP modal problems. A discussion of coupled polarization states in the end of this section contains an explanation of when the LP mode description can actually be a more accurate one than the hybrid mode description. Section three concentrates on explaining where the maxim "single mode lasers with single mode fibers, multi-mode lasers with multimode fibers" comes from, by presenting discussions of modal noise and of the coupling problem in the mode continuum and coherent coupling limits. The fourth section of the chapter is a discussion of dispersion and its relation to system performance.

## 5.1 SOME FIBER HISTORY

It is not really as if the optical fiber concept is exactly new. The dielectric cylinder problem had already been solved by Hondros and Debye (1910). Zahn (1916) fabricated microwave dielectric guides and made propagation measurements. Detailed measurements of higher order mode propagation were made by Schreiver (1920). By the 1930s, the telecommunications researchers at Bell Laboratories had adopted the problem (Southworth 1936, Carson, Mead and Schelkunoff 1936) as well as some academic researchers (Barrow 1936). World War II, however, seemed to put a damper on such research efforts, at least temporarily. Interest, however, was immediately

**145**

restimulated by Maimon's demonstration of the ruby laser in 1960 (Maimon 1960), as is indicated by the exhaustive study of fused silica optical fiber waveguides published by Snitzer (1961). A further stimulus for fiber research and development came with the almost simultaneous demonstration of the cryogenically cooled semiconductor laser by several groups in 1962 (Basov, Krokhin and Popov 1961, Hall, Fenner, Kinsley, Soltys and Carlson 1962, Nathan, Dumke, Burns, Dills and Lasher 1962). Fiber losses, however, remained very high, greater than 1 dB/meter, throughout the 1960s. It was Jones and Kao (1969), however, who noted that these losses were not inherent to the glass but were caused by impurities and that the inherent Rayleigh scattering contribution to glass loss was extremely low. It was not until after some pioneering theoretical waveguide work by Kurtz and Streifer (1969) as well as the demonstration of the room temperature semiconductor laser by several groups in 1970 (Hayashi and Parish 1970, Alferov, Andreev, Portnoi and Trukan 1970) that the first low loss (less than 100 dB/km) was demonstrated by Kapron, Keck and Mauer (1970). After this point, developments were blindingly rapid. By 1975, the first wholly semiconductor laser driven fiber optic telephone trunk lines were in the ground. By 1980, single mode fiber was the transmission medium of essentially all long haul telephone systems being installed. By 1988, the Bell system had a transatlantic fiber cable with undersea repeaters. Although this list could go on and on, really important points about the explosive growth of the optical communications field are that it was driven by a combination of the fiber and the semiconductor laser. Of course, the development of optical communications was not self-primed. The development of large digital switches in the mid-1960s clearly brought about a need for a faster, lighter, more compact transmission medium which could serve to exploit the new switching technology. That the development of fiber and laser technology took place at roughly the same time, however, allowed developments in each field to reinforce developments in the other and therefore create a self-driving technology.

## 5.2 THE MODES OF OPTICAL FIBERS

The problem of a homogeneous dielectric cylinder was given some discussion in the first section of Chapter 4. The discussion culminated in Equation (4-14), in which the solutions for $E_z$ and $H_z$ for the problem of Figure 4.4 were expressed. What remained to be done with this solution was to combine it with Equation (4-13) and use the fields of Equations (4-13) and (4-14) to satisfy the boundary conditions discussed in Chapter 2 in combination with the discussion of the Fresnel problem. The result of solving these continuity equations at $r = a$ would in turn give the dispersion relation, whose solution would give the value of $\beta$, the propagation constant, which could then be back substituted to give the constants $A_\nu$, $B_\nu$, $C_\nu$, and $D_\nu$, of Equation (4-14) as a one-parameter string. This solution could then allow one to determine the propagation properties of the fiber modes. This is the project that we will carry out in the present section.

If one uses Equations (4-13) and (4-14) to satisfy the continuity conditions on $E_z$, $\partial E_z/\partial r$, $\epsilon_i H_z$, and $\epsilon_i \partial H_z/\partial r$ at the $n_1 - n_2$ boundary at $r = a$, one finds the relations (where $l$ has been substituted for the $\nu$ of Chapter 4)

$$\left(\frac{1}{\gamma^2} + \frac{1}{\kappa^2}\right) \frac{i\beta l}{a} A_l$$

$$= -i\omega\mu_0 \left[\frac{1}{\gamma J_l(\gamma a)} \frac{dJ_l(\gamma r)}{dr}\right)\bigg|_{r=a} + \frac{1}{\kappa K_l(\kappa a)} \frac{dK_l(\kappa r)}{dr}\bigg|_{r=a}\right] B_l \quad \text{(a)}$$

$$\left(\frac{1}{\gamma^2} + \frac{1}{\kappa^2}\right) \frac{i\beta l}{a} B_l$$

$$= -i\omega \left[\frac{\epsilon_1}{\gamma J_l(\gamma a)} \frac{dJ_l(\gamma r)}{dr}\right)\bigg|_{r=a} + \frac{\epsilon_2}{\kappa K_l(\kappa a)} \frac{dK_l(\kappa r)}{dr}\bigg|_{r=a}\right] A_l \quad \text{(b)} \quad \text{(5-1)}$$

From here on we use the abbreviated notation that

$$J_l'(\gamma a) = \frac{dJ_l(\gamma r)}{dr}\right)\bigg|_{r=a} \tag{a}$$

$$K_l'(\kappa a) = \frac{dK_l(\kappa r)}{dr}\right)\bigg|_{r=a} \tag{b} \quad (5\text{-}2)$$

Taking the determinant of the coefficients of (5-1), one can immediately find that the dispersion relation is given by

$$\left[\frac{J_l'(\gamma a)}{\gamma J_l(\gamma a)} + \frac{K_l'(\kappa a)}{\kappa K_l(\kappa a)}\right]\left[\frac{J_l'(\gamma a)}{\gamma J_l(\gamma a)} + \frac{n_2^2}{n_1^2}\frac{K_l'(\kappa a)}{\kappa K_l(\kappa a)}\right] = \frac{\beta^2 l^2}{a^2}\left(\frac{1}{\gamma^2} + \frac{1}{\kappa^2}\right)^2 \tag{5-3}$$

The question now arises, naturally, as to how one solves Equation (5-3). Clearly, for the case $l = 0$ this is possible.

With $l = 0$, there are two possibilities. The first one is

$$\frac{J_0'(\gamma a)}{\gamma J_0(\gamma a)} + \frac{K_0'(\kappa a)}{\kappa K_0(\kappa a)} = 0 \tag{5-4}$$

which will yield a set of $\beta$'s which will be subscripted by the $l = 0$, $m$ values $\beta_{0m}^{\text{TE}}$. The $m$ is due to the fact that the $J_0$ has an infinite number of sign changes and therefore, for the fiber radius going to infinity, (5-4) would have a countably infinite number of solutions. The TE superscript comes from the fact that back substitution into Equations (5-1) would yield $A_0 = 0$ and therefore uncover the fact that this mode has field components $E_r$, $E_\theta$, $H_r$, $H_\theta$, and $H_z$, and is therefore a transverse electric mode.

The second possibility for $l = 0$ is that the dispersion relation is given by

$$\frac{J_0'(\gamma a)}{\gamma J_0(\gamma a)} + \frac{n_2^2}{n_1^2} \frac{K_0'(\kappa a)}{\kappa K_0(\kappa a)} = 0 \tag{5-5}$$

This system will yield a set of $\beta$'s which will be labeled $\beta_{0m}^{TM}$. The TM superscript here is due to the fact that a substitution into Equations (5-1) of Equation (5-5) would yield $B_0 = 0$. For $B_0 = 0$, the surviving field components would be $E_r$, $E_\theta$, $E_z$, $H_r$, and $H_\theta$, which are indeed those which comprise a transverse magnetic mode.

We could proceed to try to solve the dispersion relation of (5-3) in all its horror for all values of $l$. However, as discussed previously, dielectric guides are quite generally weakly guiding. This means that the $n_1$ must be given by $n_1 = n_2 + \Delta n$ where $\Delta n \ll n_1, n_2$. Use of this fact will greatly simplify the problem and, as it turns out, by "almost" equivalent to the LP mode approximation to be made in a subsequent paragraph. Expanding Equation (5-3) in $\Delta n$ and equating the zeroth order terms, one finds

$$\left[ \frac{J_l'(\gamma a)}{\gamma J_l(\gamma a)} + \frac{K_l'(\kappa a)}{\kappa K_l(\kappa a)} \right]^2 = \frac{\beta^2 l^2}{\kappa_1^2 a^2} \left( \frac{1}{\gamma^2} + \frac{1}{\kappa^2} \right)^2 \tag{5-6}$$

Applying the same expansion to Equations (5-1) yields the relation

$$\left( \frac{A_l}{B_l} \right)^2 = \frac{\mu_0}{\epsilon} \tag{5-7}$$

As we have taken $n_1 \sim n_2$, it is not immediately clear which $\epsilon$ is actually contained in (5-7). For this reason, we have the freedom to take that $\epsilon$ to be $n_{eff}^2 \epsilon_0$, where the $n_{eff}$ is given by $n_{eff} = \beta/k_0$. Using this, one finds that (5-7) is equivalent to the relation

$$E_z = \pm \eta H_z = \pm \frac{\eta_0}{n_{eff}} H_z \tag{5-8}$$

with $\eta_0 = \sqrt{\frac{\mu_0}{\epsilon_0}}$. The existence of an impedance relation such as (5-8) is reminiscent of a plane wave and, indeed, with the weakly guiding approximation the modes become plane-wavelike.

For the sake of brevity we just state a procedure for further simplifying the dispersion relation (5-6) and then discuss the results. By applying the well-known Bessel function identities

$$J_l' = \mp \frac{lJ_l}{\gamma a} \pm J_{l\mp1} \qquad \text{(a)}$$

$$K_l' = \mp \frac{lK_l}{\kappa a} + K_{l\mp1} \qquad \text{(b)} \quad \text{(5-9)}$$

one finds that the so-called $\mathbf{EH}_{lm}$ modes (those which correspond to the plus sign of Equation (5-8)) are defined by the dispersion relation

$$\frac{J_{l+1}}{\gamma J_l} - \frac{K_{l+1}}{\kappa K_l} = 0, \qquad l \geq 1 \qquad (5\text{-}10)$$

and that the so-called $\mathbf{HE}_{lm}$ modes (those which correspond to the minus sign of Equation (5-8)) are defined by the dispersion relation

$$\frac{J_{l-1}}{\gamma J_l} - \frac{K_{l-1}}{\kappa K_l} = 0, \qquad l \geq 1 \qquad (5\text{-}11)$$

The results of the dispersion relations of (5-6) for TE and TM and of (5-10) for EH and (5-11) for HE can be summed up for the first few modes by the $\beta$ line of Figure 5.1, where the modes are labelled by the hybrid mode names above the relative positions of their $\beta$ values. We will now discuss the transverse field configurations of the modes along with their degeneracy.

As the first two modes on our $\beta$ line are HE modes, discussion will first be given to the transverse field structure of these modes. Following from Equation (5-8), the longitudinal inside fields of an HE mode can be expressed in

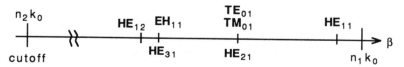

**FIGURE 5.1.**    $\beta$ line where a mode's position gives its distance from the low order cutoff point $n_1 k_0$.

terms of a single electric field amplitude $E_0$ by

$$E_z(r,\theta) = E_0 J_l(\gamma r)e^{\pm il\theta} \qquad \text{(a)}$$

$$H_z(r,\theta) = -i\frac{E_0}{\eta_{lm}} J_l(\gamma r)e^{\pm il\theta} \qquad \text{(b)} \quad \text{(5-12)}$$

where the plus or minus in the exponent is due to the fact that only $l^2$ showed up in the defining equation for the Bessel function. The plus or minus in front of the $H_z$ expression is to fix the phase relation between $E_z$ and $H_z$ such that all the terms in Chapter 4, Equation (4-13) have the same signs, which is strictly a convention for HE modes. Evidently if one had taken the minus in Equation (4-8), that is, the signs corresponding to the EH modes, all signs in (4-13) would alternate. $\eta_{lm}$ is the impedance of the $lm^{\text{th}}$ mode as defined by Equation (5-8), with $n_{\text{eff}} = \beta_{lm}/k_0$. If one plugs these dependences into Equation (4-13) of Chapter 4 for the transverse fields in terms of the longitudinal fields, one finds for $E_r$ and $E_\theta$

$$E_r = \frac{i\beta E_0 e^{\pm il\theta}}{k_i^2 - \beta^2}\left[\gamma J_l'(\gamma r) \pm \frac{l}{r} J_l(\gamma r)\right] \qquad \text{(a)}$$

$$E_\theta = -\frac{\beta E_0 e^{\pm il\theta}}{k_i^2 - \beta^2}\left[\gamma J_l'(\gamma r) \pm \frac{l}{r} J_l(\gamma r)\right] \qquad \text{(b)} \quad \text{(5-13)}$$

which can be combined to yield

$$\mathbf{E}_t = E_r\hat{\mathbf{e}}_r + E_\theta\hat{\mathbf{e}}_\theta = if_{lm}(r)e^{\pm il\theta}(\hat{\mathbf{e}}_r \mp i\hat{\mathbf{e}}_\theta) \qquad \text{(5-14)}$$

where the $f_{lm}(r)$ can be considered to be defined by the relation between Equations (5-13) and (5-14). Using the relations that

$$\hat{\mathbf{e}}_r = \hat{\mathbf{e}}_y \sin\theta + \hat{\mathbf{e}}_x \cos\theta \qquad \text{(a)}$$

$$\hat{\mathbf{e}}_\theta = -\hat{\mathbf{e}}_x \sin\theta + \hat{\mathbf{e}}_y \cos\theta \qquad \text{(b)} \quad \text{(5-15)}$$

one can show that

$$\mathbf{E}_t = if_{lm}(r)e^{\pm i(l-1)\theta}(\hat{\mathbf{e}}_x \pm i\hat{\mathbf{e}}_y) \qquad \text{(5-16)}$$

Before proceeding, let us take a moment to discuss what relation (5-16) means. The plus or minus signs mean that, for a given $lm$, there are two possible solutions, expressible as

$$\mathbf{E}_t^+ = i f_{lm}(r) e^{i(l-1)\theta} (\hat{\mathbf{e}}_x + i\hat{\mathbf{e}}_y) \qquad \text{(a)}$$

$$\mathbf{E}_t^- = i f_{lm}(r) e^{-i(l-1)\theta} (\hat{\mathbf{e}}_x - i\hat{\mathbf{e}}_y) \qquad \text{(b)} \quad \text{(5-17)}$$

As these two configurations both represent modes with the same propagation constant (degenerate modes), the total transverse field configuration for that given propagation constant would be given as

$$\mathbf{E}_t = a^- \mathbf{E}_t^- + a^+ \mathbf{E}_t^+ \qquad (5\text{-}18)$$

where the $a^-$ and $a^+$ are modal excitation coefficients, as will be discussed further in the next section.

One can make some conclusions as to the spatial behavior of the HE modes even without a detailed knowledge of the behavior of $f_{lm}(r)$. For example, Equation (5-16) tells us immediately that the $\mathbf{HE}_{1m}$ modes will have no $\theta$ dependence. We further could surmise from our earlier exposure to the modes of slab structures that the $\mathbf{HE}_{11}$ would have no radial nulls, the $\mathbf{HE}_{12}$ would have one radial null, etc. Further, the $\hat{\mathbf{e}}_x \pm i\hat{\mathbf{e}}_y$ polarization behavior we recognize as being just that of uniformly polarized plane waves. This knowledge of the $\mathbf{HE}_{1m}$ modes we sum up in the sketches of Figure 5.2. Perhaps the other most important salient feature of these $\mathbf{HE}_{1m}$ modes is that, for each $m$ value, there are actually two modes. That is, in each of the sketches of Figure 5.2, each of the arrows could have been reversed, and the result would have been an orthogonally polarized mode with the same principle mode numbers.

To find the transverse field behavior for the $\mathbf{HE}_{21}$ modes and, in general, the $\mathbf{HE}_{lm}$ modes (there are two for each $l,m$), it will be easiest to find the so-called even and odd components of the two. As the two polarization states will be degenerate, we can also rewrite the modes as linear combinations of other orthogonal states. For example, the above discussion has made clear

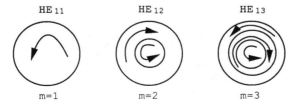

**FIGURE 5.2.**    Transverse mode structure of the three first $\mathbf{HE}_{1m}$ modes, where the arrowed lines indicate polarization direction in that given mode domain.

that $\mathbf{HE}_{lm}$, the transverse field structure of the $lm^{\text{th}}$ HE mode, can be expressed as

$$\mathbf{HE}_{lm} = f_{lm}(r)[a_{\overline{lm}}^- e^{i(l-1)\theta} (\hat{\mathbf{e}}_x + i\hat{\mathbf{e}}_y) + a_{lm}^+ e^{-i(l-1)\theta} (\hat{\mathbf{e}}_x + i\hat{\mathbf{e}}_y)] \quad (5\text{-}19)$$

where the $a_{\overline{lm}}^-$ and $a_{lm}^+$ are excitation coefficients for these two orthogonal modes. But again, as these modes are degenerate, we could just as well express the general $\mathbf{HE}_{lm}$ modes as

$$\mathbf{HE}_{lm} = f_{lm}(r)[a_{lm}^e(\cos (l-1)\theta \, \hat{\mathbf{e}}_x - \sin (l-1)\theta \, \hat{\mathbf{e}}_y) \\ + a_{lm}^0(\sin (l-1)\theta \, \hat{\mathbf{e}}_x + \cos (l-1)\theta \, \hat{\mathbf{e}}_y)] \quad (5\text{-}20)$$

where the motivation for the even and odd designation is that the even modes will have cos multiplying $\hat{\mathbf{e}}_x$ and odd modes will have cos multiplying $\hat{\mathbf{e}}_y$. The $\mathbf{HE}_{21}$ modes, even and odd, are sketched in Figure 5.3.

The next two modes on our $\beta$ line of Figure 5.1 are the $\mathbf{TE}_{01}$ and $\mathbf{TM}_{01}$, which are degenerate with the $\mathbf{HE}_{21}$. A determination of the salient features of the transverse mode structures of these modes is more straightforward than it was for the higher order HE's. For example, for a TE mode, the $E_z$ is zero and, from the condition that $l = 0$, one must also have $\partial H_z/\partial\theta = 0$. Putting these conditions into Equation (5-12) immediately gives that $\mathbf{TE}_{0m}$, the transverse mode structure of the $m^{\text{th}}$ TE mode, must be of the form

$$\mathbf{TE}_{0m} = f_{0m}(r)\hat{\mathbf{e}}_\theta \quad (5\text{-}21)$$

which indicates that there is only one $\mathbf{TE}_{0m}$ polarization configuration. That there were two $\mathbf{HE}_{lm}$ polarization configurations stemmed from the fact that one could take $\pm l$ in the original equations. Whem $l = 0$, there is only one choice for $l$ and a single polarization state is possible. This single polarization state is sketched for the first three TE modes in Figure 5.4.

The TM mode configurations are derivable as they were for the TE. One

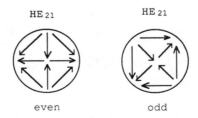

**FIGURE 5.3.**     Transverse mode configuration for the even and odd $\mathbf{HE}_{21}$ modes.

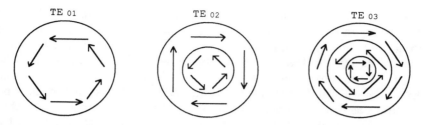

**FIGURE 5.4.**    Transverse mode configurations for the first three transverse electric modes.

first notes that, for TM, $H_z = 0$ and $\partial E_z/\partial\theta = 0$ to find that

$$\mathbf{TM}_{0m} = g_{0m}(r)\hat{\mathbf{e}}_r \tag{5-22}$$

where $\mathbf{TM}_{0m}$ is the transverse mode configuration and $g_{0m}(r)$ is a radial mode variation. Again, as was the case with the $\mathbf{TE}_{0m}$, there is but one $\mathbf{TM}_{0m}$. The first three of these transverse mode configurations are sketched in Figure 5.5.

The third group of modes on our $\beta$ line consists of the $\mathbf{EH}_{11}$ and the $\mathbf{HE}_{31}$, and not too far away in $\beta$ distance is the $\mathbf{HE}_{12}$. From the above discussions, we have a very good idea about the structures of the HE modes. The $\mathbf{HE}_{31}$ will vary with sinusoids of $2\theta$ but will have no radial nulls. The $\mathbf{HE}_{12}$ will be azimuthally symmetric but will possess a radial null. Each of these modes will have two distinct polarization configurations, as will the $\mathbf{EH}_{11}$. Further, it is obvious that the $\mathbf{EH}_{11}$ and $\mathbf{HE}_{31}$ must be degenerate as, indeed, the dispersion relations of Equations (5-10) and (5-11) show that $\mathbf{HE}_{j+1,m}$ is always degenerate with $\mathbf{EH}_{j-1,m}$. As we have already been through the derivation of the transverse mode structure of the HE's and we know that the only difference in the derivation for the EH's will be the sign in Equation (5-8), we can just as well write down the result

$$\mathbf{EH}_{lm} = g_{lm}(r)[a_{lm}^e(\cos(l+1)\theta\,\hat{\mathbf{e}}_x + \sin(l+1)\theta\,\hat{\mathbf{e}}_y)$$
$$+ a_{lm}^0(\sin(l+1)\theta\,\hat{\mathbf{e}}_x - \cos(l+1)\theta\,\hat{\mathbf{e}}_y] \tag{5-23}$$

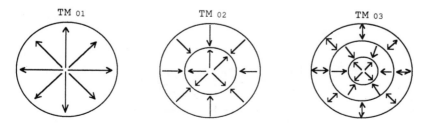

**FIGURE 5.5.**    Transverse mode configurations for the first three transverse electric modes.

Essentially what we have done is to go through a rather complicated derivation, find a complicated answer, and then start simplifying it until we got to something we could "kind of" handle. The polarization states, however, of all but the $HE_{11}$ mode are things we are rather unused to. The question therefore arises as to what would happen were we to start the derivation with all the possible simplifications and then solve the resulting equations as exactly as possible. How will this result compare with that derived? This is the subject of the next few paragraphs in which the LP (linearly polarized) modes of a parabolic index fiber will be derived.

What we wish to do is to consider the form that Maxwell's equations would take for an index profile (the $\alpha$ profile) of the form

$$n^2(r) = n_1^2 \begin{cases} 1 - 2\Delta \left(\dfrac{r}{a}\right)^{\alpha}, & r < a \\ 1 - 2\Delta, & r > a \end{cases} \tag{5-24}$$

where the parameter $\Delta$ can be taken to be small, $\Delta \ll 1$. This index profile is sketched in Figure 5.6 for several values of $\alpha$. The $\alpha$ value of infinity (step index) would correspond to that of our hybrid modes, where the $n_2 = n_1 \sqrt{1 - 2\Delta}$. In the present discussion, we shall solve the wave equation only for $\alpha = 2$, but we shall later see explicitly that the solutions are not qualitatively changed for $\alpha$ values ranging from $\alpha = 1$ to $\infty$, and therefore can still make comparison between our LP solutions and our hybrid solutions.

Assuming a solution of the form

$$E(r,t) = \text{Re}[E(r)e^{-i\omega t}] \tag{5-25}$$

and assuming a spatially varying index in Maxwell's equations, one finds the vector equation for the $E$-field

$$\nabla^2 E + k^2 E = -\nabla \left[ \frac{E \cdot \nabla \epsilon}{\epsilon} \right] \tag{5-26}$$

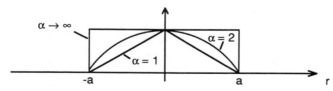

**FIGURE 5.6.**    Profile function of Equation (5-24) for several values of the $\alpha$ parameter.

However, have we already assumed that $\Delta$ is small and that we could there-fore set up a perturbation expansion in the small parameter $\Delta$. As has been mentioned previously in these notes, the term on the right-hand side repre-sents the coupling of the polarization states. Further, the forms of (5-26) and (5-24) indicate that the polarization term will be multiplied by $\Delta$. If one were to assume a field solution of the form

$$\mathbf{E}(\mathbf{r}) = \sum_{n=0}^{\infty} \Delta^n \mathbf{E}^{(n)}(\mathbf{r}) \tag{5-27}$$

plug this into (5-26) (assume that $k^2(r)$ has no $\Delta$ in it), and equate successive terms in $\Delta$ to obtain equations for the $\mathbf{E}^{(n)}(\mathbf{r})$. If we carried this out, we would find that the zeroth order term would satisfy

$$\nabla^2 \mathbf{E} + k^2 \mathbf{E} = 0 \tag{5-28}$$

as the term $\nabla \epsilon$ contains $\Delta$ to first order. As in the hybrid solution, we completely ignored the first order corrections in $\Delta$. We shall do the same here and consider Equation (5-28) to be the "correct" equation. A very important thing about Equation (5-28) is that it splits up into three uncoupled scalar equations for the Cartesian components of the wave field. We can now see that our $\Delta$ expansion has already dictated that the fiber modes will be linearly polarized (LP) and that a general solution for the transverse field configura-tion could be expressed in the form

$$\mathbf{E}_t(\mathbf{r},t) = a_x E_x(\mathbf{r},t)\hat{\mathbf{e}}_x + a_y E_y(\mathbf{r},t)\hat{\mathbf{e}}_y \tag{5-29}$$

Assuming that the $E_x$ and $E_y$ of Equation (5-29) are expressible in the forms

$$E_x(r,t) = \mathrm{Re}[\psi(r)e^{iv\theta}e^{i\beta z}e^{-i\omega t}] \tag{a}$$

$$E_y(r,t) = \mathrm{Re}[\psi(r)e^{iv\theta}e^{i\beta z}e^{-i\omega t}] \tag{b} \tag{5-30}$$

where $r$ and $\theta$ are the radial and azimuthal coordinates of a polar coordinate system, then Maxwell's equations reduce to the form

$$\frac{d^2\psi(r)}{dr^2} + \frac{1}{r}\frac{d\psi}{dr} + \left(k^2(r) - \beta^2 - \frac{v^2}{r^2}\right)\psi = 0 \tag{5-31}$$

where the $k^2(r)$ is given by

$$k^2(r) = k_1^2\left[1 - 2\Delta\left(\frac{r}{a}\right)^\alpha\right] = k_0^2 n_1^2\left[1 - 2\Delta\left(\frac{r}{a}\right)^\alpha\right] \tag{5-32}$$

One might note at this point that the perturbation expansion we did to (5-26) to obtain (5-28) was not especially consistent if we can turn around and stick a $\Delta$ back into the $k^2(r)$. The point is, however, that $k^2$ is big (in terms of inverse wavelengths), and this multiplies the $\Delta$. This is to say that the $\Delta$ can have a profound effect on the wave's phase and, therefore, $\beta$ value. The expansion is not really completely consistent, but we shall see later that it gives reasonable answers. Anyway, if one were to drop the $\Delta$ in (5-32), there would be no guidance, and the problem would be totally uninteresting.

We now wish to specialize to the case of $\alpha = 2$. Through a series of transformations and an additional approximation, we will obtain an explicit relation for the propagation constants of this so-called parabolic index profile. The equation is at present of the form

$$\frac{d^2\psi}{dr^2} + \frac{1}{r}\frac{d\psi}{dr} + \left(k_1^2 - \beta^2 - \frac{2\Delta k_1^2 r^2}{a^2} - \frac{v^2}{r^2}\right)\psi = 0 \qquad (5\text{-}33)$$

Performing the transformation $\xi = cr^2$, one can obtain the form

$$\xi\frac{d^2\psi}{d\xi^2} + \frac{d\psi}{d\xi} + \frac{1}{4}\left(\lambda - \xi - \frac{v^2}{\xi}\right)\psi = 0 \qquad (5\text{-}34)$$

where $c$ and $\xi$ are defined as

$$c^2 = \frac{2\Delta k_1^2}{a^2} \qquad\qquad\qquad\text{(a)}$$

$$\lambda = \frac{k_1^2 - \beta^2}{c} \qquad\qquad\text{(b)} \quad (5\text{-}35)$$

By making the substitution $\psi = \xi^{v/2}\phi$, one can show that

$$\xi\frac{d^2\phi}{d\xi^2} + (1 + v)\frac{d\phi}{d\xi} + \frac{1}{4}(\lambda - \xi)\phi = 0 \qquad (5\text{-}36)$$

This equation can be transformed into the form of the confluent hypergeometric equation through use of the transformation $\phi = e^{-\xi/2}L$ to yield

$$\xi\frac{d^2L}{d\xi^2} + (1 + v - \xi)\frac{dL}{d\xi} + \frac{1}{4}(\lambda - 2(1 + v))L = 0 \qquad (5\text{-}37)$$

At this point, we should pause to consider where we are and where we are going. Equation (5-37) is for the field inside the guide. We note that the

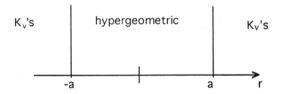

**FIGURE 5.7.** Boundary value problem that would need to be solved to determine the $\lambda$ value of Equation (5-37).

equation for the field outside the guide's core will be satisfied by the $Kv$'s, the modified Bessel functions that we have seen repeatedly before. The situation is as depicted in Figure 5.7. One could write down the solution of Equation (5-37) with $\lambda$ as a parameter and then equate the inside and outside tangential fields around the cylindrical surface at $r = a$ to obtain a dispersion relation. This would not be especially easy to do, as we are using linear polarization states and they do not fit nicely to the boundary. That is, the $E_z$ and $H_z$ will remain tangential, but different combinations of the $E_x$, $E_y$, $H_x$, and $H_y$ would be needed at different points to satisfy the boundary conditions. It would be nice if we could somehow ignore the boundary conditions. Figure 5.8 seems to show us that we can, at least for the lower order modes of a very multimode waveguide. The idea is that, in a multimode guide, there will be a number of modes which are so well bound that they really can not see the core cladding interface. If they do not see this boundary, it should make no real difference if we allow the profile to extend to infinity as in Figure 5.9. In the next paragraph, we shall see why this approximation is of great utility.

An equation such as (5-37) should have a series solution of the form

$$L(\xi) = \sum_{n=0}^{\infty} a_n \xi^n \tag{5-38}$$

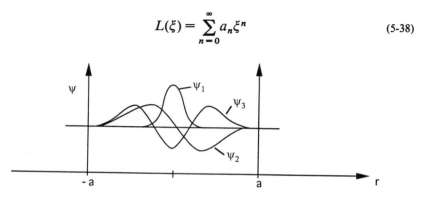

**FIGURE 5.8.** The first three modes of a multimode waveguide.

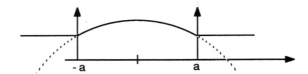

**FIGURE 5.9.**    The effect of the approximation of the infinite profile.

As the profile extends to $\xi \to \infty$, there are some restrictions on the $a_n$'s. For example, the recursion relation obtained for the $a_n$'s by substituting (5-38) into (5-37) should be truncated at some finite $n$, lest our mode grow without bound for $\xi \to \infty$. This would not be modelike behavior. The coefficients in (5-37) are combinations of constants and integral (namely, first) powers of $\xi$. These first power of $\xi$ will couple adjacent $\xi_n$'s. It should not be difficult to convince oneself that were the coefficient of $L$ in (5-37) not integer the recursion for $a_n$ could not terminate, as at each step we would be equating integer and noninteger coefficients. It is not hard either to show that the series terminates for all integer values of the coefficient of $L$ and, therefore, that the condition for the existence of a mode becomes

$$\frac{1}{4}[\lambda - 2(1+v)] = \mu = \text{integer} \tag{5-39}$$

Using relations (5-35) in (5-39), one immediately obtains

$$\beta_{\mu v} = k_1 \left[ 1 - 2\frac{\sqrt{2\Delta}}{k_1 a}(2\mu + v + 1) \right]^{1/2} \sim k_1 \left[ 1 - \frac{\sqrt{2\Delta}}{k_1 a}(2\mu + v + 1) \right] \tag{5-40}$$

where the last part of (5-40) required a reuse of the weakly guiding condition $2\Delta \ll 1$. The result of this equation is summed up on the $\beta$ line of Figure 5.10. Note that the values of $\beta$ are approximately evenly spaced.

To get explicit expressions for the LP modes, we need explicit solutions

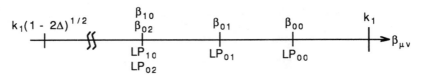

**FIGURE 5.10.**    $\beta$ line for Equation (5-40).

for the equations

$$\xi \frac{\partial^2 L}{\partial \xi^2} + (1 + v - \xi) \frac{dL}{\partial \xi} + \mu L = 0 \qquad (5\text{-}41)$$

which can serve as a defining equation for the Laguerre polynomials $L_\mu^v(\xi)$. The first few of these are given by

$$L_0^0 = 1 \qquad \text{(a)}$$

$$L_0^1 = 1 \qquad \text{(b)}$$

$$L_0^2 = 2 \qquad \text{(c)}$$

$$L_0^3 = 3! \qquad \text{(d)}$$

$$L_0^m = m! \qquad \text{(e)}$$

$$L_1^0 = 1 - \xi \qquad \text{(f)}$$

$$L_1^1 = 4 - 2\xi \qquad \text{(g)}$$

$$L_2^0 = 2 - 4\xi + \xi^2 \qquad \text{(h)} \quad (5\text{-}42)$$

Using (5-42) and the fact that the fiber V number $V = \sqrt{2\Delta} k_1 a$, one can use the substitutions of (5-33)–(5-37) together with (5-30) and (5-31) to find the expression for $\psi_{\mu v}(x,y,z)$ of the LP modes as

$$\psi_{\mu v} = N_{\mu v} r^v e^{-\frac{Vr^2}{2a^2}} L_\mu^v \left( \frac{Vr^2}{a^2} \right) e^{iv\theta} e^{i\beta_{\mu v} z} \qquad (5\text{-}43)$$

where $N_{\mu v}$ is a normalization factor that we will not bother to calculate. Specifically, we can write out the first few modes as

$$\psi_{00} = N_{00} e^{-\frac{Vr^2}{2a^2}} e^{i\beta_{00} z} \qquad \text{(a)}$$

$$\psi_{01} = N_{01} r e^{-\frac{Vr^2}{2a^2}} e^{\pm i\theta} e^{i\beta_{01} z} \qquad \text{(b)}$$

$$\psi_{02} = N_{02} r^2 e^{-\frac{Vr^2}{2a^2}} e^{\pm 2i\theta} e^{i\beta_{02} z} \qquad \text{(c)}$$

$$\psi_{10} = N_{10} \left( 1 - \frac{Vr^2}{a^2} \right) e^{-\frac{Vr^2}{2a^2}} e^{i\beta_{10} z} \qquad \text{(d)} \quad (5\text{-}44)$$

where the radial parts of these four functions are sketched in Figure 5.11.

It is now necessary to discuss the modal polarizations, much as we did for the first few hybrid modes. Clearly, there will be a difference between the

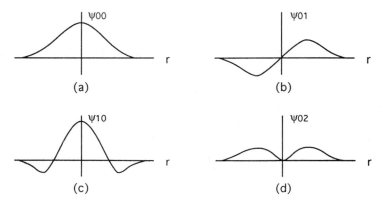

**FIGURE 5.11.**     Radial variations of the first few LP modes.

$LP_{\mu v=0}$ and the $LP_{\mu v \neq 0}$ modes, as the $LP_{\mu 0}$ modes have no azimuthal dependence. These modes, however, will have two polarization states, $E_x$ and $E_y$. A general $LP_{\mu 0}$ state can therefore be expressed in the form

$$\mathbf{E}_{\mu 0} = a_{\mu 0 x} \psi_{00} \hat{\mathbf{e}}_x + a_{\mu 0 y} \psi_{00} \hat{\mathbf{e}}_y \tag{5-45}$$

where the $a_{\mu 0 x}$ and $a_{\mu 0 y}$ are coefficients to be determined from the excitation conditions, as will be done in a subsequent section. The $\mu v$ modes are actually going to each be comprised of four modes, two polarizations each of two orientational modes. That is, the general vector expression for an $LP_{\mu v}$ mode would look like

$$\mathbf{E}_{\mu v} = \psi_{\mu v}(r,z)[a^+_{\mu v x} e^{iv\theta} \hat{\mathbf{e}}_x + a^+_{\mu v y} e^{iv\theta} \hat{\mathbf{e}}_y + a^-_{\mu v x} e^{-iv\theta} \hat{\mathbf{e}}_x + a^-_{\mu v y} e^{-iv\theta} \hat{\mathbf{e}}_y \tag{5-46}$$

where the $\psi_{\mu v}(r,z)$ is just $\psi_{\mu v}(r,\theta,z)$ of Equations (5-44) with the $v\theta$ variation factored out and $a^+_{\mu v x}$, $a^+_{\mu v y}$, $a^-_{\mu v x}$, and $a^-_{\mu v y}$ are excitation coefficients. By analogy with Equation (5-20), one could just as well have written $\mathbf{E}_{\mu v}$ in terms of even and odd parts as

$$\mathbf{E}_{\mu v} = \psi_{\mu v}(r,z)[a^e_{\mu v x} \cos v\theta \, \hat{\mathbf{e}}_x + a^e_{\mu v y} \cos v\theta \, \hat{\mathbf{e}}_y + a^0_{\mu v x} \sin v\theta \, \hat{\mathbf{e}}_x + a^0_{\mu v y} \sin v\theta \, \hat{\mathbf{e}}_y] \tag{5-47}$$

where the $a$'s are excitation coefficients. The results of equations (5-45) and (5-47) are summed up in Figure 5.12, where each given eigenpolarization state is plotted.

As we have obtained some knowledge of the weakly guiding hybrid and LP modes, an interesting problem to attack is that of the relationship be-

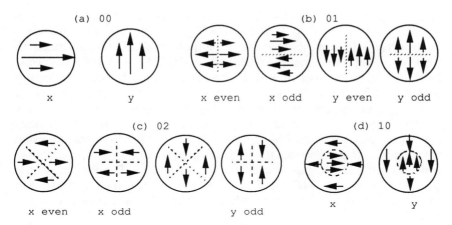

**FIGURE 5.12.** Polarization of the modes comprising a given $\mu\nu$ mode number for the first few modes.

tween them. As their respective derivations entailed repeated use of the same approximations, one would think that, at least for those modes that do not see the boundary too much, the two sets of solutions should correspond closely to each other. One would further think that, at least qualitatively, all the bound modes should exhibit some agreement. We shall soon see that this is indeed the case. There is already a strong hint that the two sets of modes are closely related, which one can see from plotting the $\beta$ lines for the two mode systems one on top of the other, as in Figure 5.13. It is clear from the figure that somehow the $LP_{00}$ should correspond to the $HE_{11}$, the $LP_{01}$ to a combination of the $HE_{21}$, $TM_1$, and $TE_1$, the $LP_{02}$ to a combination of the $HE_{31}$

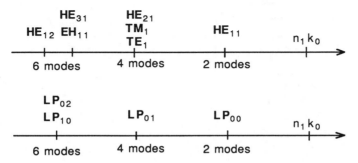

**FIGURE 5.13.** Locations of the first few hybrid modes and LP modes on equivalent $\beta$ lines, with the number of degenerate modes at a given $\beta$ value written in below that $\beta$ value.

and $EH_{11}$, and the $LP_{10}$ to the $HE_{12}$. That there is a slight splitting between the $HE_{31}$ and $HE_{12}$ and not between the $LP_{02}$ and $LP_{10}$ should not disturb us too much as it turns out that the splitting is quite small. Had we not made the weakly guiding approximation repeatedly to the $HE_{21}$ and $TM_1$, we would have found a splitting there as well. The neglection of the boundary condition in the LP solution, as well as the LP solution being for $\alpha = 2$ instead of $\alpha \to \infty$, can easily account for this splitting. The important thing is that the splitting is small and that comparably small splitting would be seen breaking the perfect degeneracies of higher order modes also. As will be discussed below, it is quite possible that the stress induced birefringence of "real" fiber will cause a larger splitting between $x$ and $y$ states of an LP mode than would be observed between $HE_{31}$ and $HE_{12}$ anyway, and certainly larger than would be observed in an exact calculation of the splitting between $HE_{21}$ and $TM_1$.

Figure 5.14 illustrates the polarization diagrams for the $x$ and $y$ $LP_{00}$ modes next to those for the $HE_{11}$ even and $HE_{11}$ odd modes. As is evident from the diagram, the mode configurations are identical in this case. Therefore, if one were to consider a single mode fiber in which a sizable portion of the energy could be propagating outside of the core, it would probably be better to use a hybrid approach to predicting the modal properties. In other words, it would be best to expand in the cylindrical coordinates, integrate numerically in the core region for nonstep profiles, taking into account any intrinsic birefringence, and match boundary conditions to find the $\beta$'s. As the states will come out to be linear, the effect of the birefringence for this fundamental mode should be calculable. For higher modes, it probably

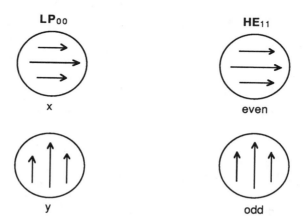

**FIGURE 5.14.** Polarization profiles for the two degenerate modes of the lowest order mode groups of both the LP and corresponding hybrid group.

would be very hard to take into account. But were there higher order modes, then there should be some number of modes which are not strongly affected by the cladding and an LP description should be not only sufficient but, perhaps, preferable for higher order modes, as birefringence can be easily taken into account. We shall see later that the so-called WKB method will allow us to treat LP modes for arbitrary $\alpha$ profiles.

In Figure 5.15, the polarization profiles for the **LP$_{01}$** and **HE$_{21}$**, **TE$_1$**, **TM$_1$** hybrid combination are illustrated. As is evident from the figure, these profiles are nonidentical. However, if one stares long enough at the modes, it becomes obvious that the **LP$_{01}$** $x$ even is just a sum of **HE$_{21}$** even and the **TM$_1$** as was noted by Snyder and Young (1978). Further, the **LP$_{01}$** $x$ odd is just the difference of the **TE$_1$** and the **HE$_{21}$** odd, the **LP$_{01}$** $y$ even is just the

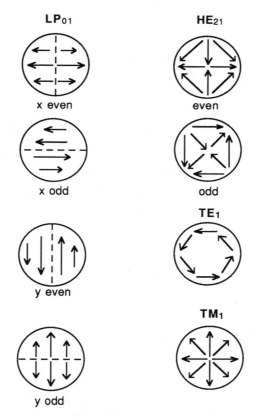

**FIGURE 5.15.** Polarization profiles for the four degenerate modes of the second mode groups of both the **LP$_{01}$** and corresponding hybrid cluster.

sum of the **TE**$_1$ and the **HE**$_{21}$ odd, and the **LP**$_{01}$ $y$ odd is just the difference of the **TM**$_1$ and the **HE**$_{21}$ even. Were a fiber strongly guiding and perfectly circular, there would be a splitting between the **TM**$_1$ and the **HE**$_{21}$ even, and if a **LP**$_{01}$ $x$ even were excited, it would not remain an **LP**$_{01}$ $x$ even with propagation distance but would evolve as the **HE**$_{21}$ even and **TM**$_1$ dephase and rephase again. However, it is clear that we could also express the **HE**$_{21}$ even as a difference of the **LP**$_{01}$ $x$ even and the **LP**$_{01}$ $y$ odd. Were the fiber to be slightly birefringent, then an **HE**$_{21}$ even excited would not remain an **HE**$_{21}$ even but would evolve as it propagated down the fiber and the **LP**$_{01}$ $x$ even and **LP**$_{01}$ $y$ odd phased and dephased with distance. It is hoped that the

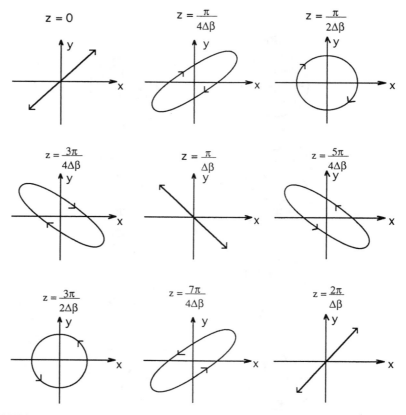

**FIGURE 5.16.**    Evolution of the polarization in a birefringent fiber which is excited in a linear polarization state at 45 degrees to the principal axes, where each separate plot is a time evolution of the tip of the polarization vector with axes $x = a_x \cos \omega t$ and $y = a_y \cos (\omega t + \Delta\beta z)$, as in a Lissajous pattern.

ramifications of this point will become clearer with the following discussion of propagation in weakly guiding birefringent fibers.

Let us first consider propagation in an ideal single mode birefringent fiber—that is, one that has fixed axes at each point along the fiber. At the input plane to the fiber, $z = 0$, let us say that the transverse electric field

$$\mathbf{E}_t = a_x f_0(r)\hat{\mathbf{e}}_x + a_y f_0(r)\hat{\mathbf{e}}_y \tag{5-48}$$

is excited, where it has tacitly been assumed that the birefringence is of a type that splits the propagation constants of the two polarization states without very strongly modifying the field distribution. At an arbitrary plane $z$ in the fiber, the transverse field will be given by

$$\mathbf{E}_t = f_0(r)e^{i\beta_x z}[a_x \hat{\mathbf{e}}_x + a_y \hat{\mathbf{e}}_y e^{-i\Delta\beta z}] \tag{5-49}$$

where $\Delta\beta = \beta_x - \beta_y$. If equal amounts of $a_x$ and $a_y$ were initially excited, then the polarization state of the wave would evolve with propagation distance in the manner depicted in Figure 5.16. The point in the figure is that the phases of the two polarization modes vary as $\Delta\beta z$ varies from 0 to $2\pi$, causing an unstable polarization state. However, in the present example of a "perfect" birefringent fiber, we see that this instability can be avoided by exciting the fiber along a principal axis—that is, by exciting only the $a_x$ or only the $a_y$. As we shall see in Section 5.4, this is only true for large birefringence.

## 5.3 SOME COMMENTS ON FIBER FABRICATION

Before we discuss real fibers, it is appropriate to devote some space to a discussion of a representative fiber fabrication process. The technique discussed here will be the modified chemical vapor deposition (MCVD) process by which fiber cores are grown and a pulling process by which a preform becomes a fiber. This is by no means the only fabrication process in use, but is the one that was used by the research group in Trondheim of which the author was a member in the 1980s. The discussion here will not be fully general and in fact will be quite cursory as to fabrication details. It is hoped that it will serve to illustrate some of the effects encountered in fiber fabrication. Interestingly enough, many of the imperfections generated during this fabrication process are also generated by other fabrication techniques, although in different degrees depending on the details of the process.

The MCVD fiber fabrication process begins with the purchase of a hollow glass tube roughly a meter in length and some centimeters in diameter. This piece is purchased because this material will form the outer parts of the cladding, which are more for mechanical stability than any propagation effect, as essentially no light will be guided in this region. There is no reason, therefore, for this material to be anywhere near as pure as the core material. There is an argument, however, for this material to be as cheap as possible in order to keep the price of the completed fiber as low as possible. To form the purest possible inner cladding and core layers, it is necessary to actually grow these layers in place by some epitaxial method, commonly MCVD. A possible setup for carrying out MCVD is schematically depicted in Figure 5.17. To keep the material as defect free as possible, it is necessary to grow a material on the interior of the tube that is as close in composition to the fused silica tube as possible yet that will still have the required difference in index. The main carrier gas flowing down the tube will therefore be a silicon derivative such as silane. Commonly used dopants are germanium (to raise the index) and boron (to lower the index). Gases containing these dopants must also be injected into the main flow down the tube. Were there no heat zone in the tube, the mixture of gases would likely just flow out the other end of the tube, and there would be no process. The moving heat zone, however, forms the zone in which the flows react. Due to symmetry, the temperature and reaction rates will be maximized in the center of the tube. Due to the higher temperature, the collision rate there is higher than that nearer the inner tube walls, therefore causing a net flow of material toward the preform's inner walls, as depicted in Figure 5.18. By slowly moving the heat zone down the tube while rotating the tube at a uniform rate, one can generate a very uniform layer of doped material on the interior of the preform. By programming the gas flows to change with each successive layer, one can control the index profile reasonably well. Small gas flow instabilities can cause perturbations to the profile, and these perturbations will have a tendency to be random, although clearly slowly varying compared to a wavelength. Even

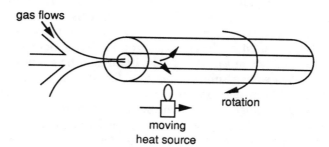

**FIGURE 5.17.**    MCVD setup for sintering chemical reactants onto the interior surface of a hollow glass tube.

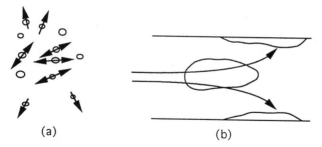

(a)                                    (b)

**FIGURE 5.18.**    The process which drives the reacted material to stick to the preform's inner walls.

though these index perturbations may be small in absolute magnitude ($\Delta n \leq 10^{-3}$ whereas the maximum index at the core center may be 0.03 for a single-mode fiber or as much as 0.1 for a multimode fiber) they still may have a measurable effect, especially on the temporal variation of light Rayleigh backscattered out of a forward propagating pulse of light (Eriksrud and Mickelson 1982b, Mickelson and Eriksrud 1982, Eriksrud and Mickelson 1982a, Mickelson and Eriksrud 1981). There are other effects that can be noticed, however, especially in multimode gradient index fiber. On the one hand in a graded index fiber, the index decreases with increasing radius. An absolutely small index perturbation near the core cladding boundary may therefore be relatively large and could cause significant mode dependent attenuation. Also, even small index changes can have a significant effect on the group velocity as a function of mode number profile, thereby significantly affecting the fiber bandwidth (Yadlowsky 1992).

After forming the fiber core in the preform, one wishes to pull out a several kilometer long, 125 $\mu$m outer diameter fiber from the roughly meter long, several centimeter diameter preform. This process is generally performed by a drawing tower as depicted in Figure 5.19. Here, by pulling the molten preform out of a furnace at a much higher velocity ($V_2$) than it is being fed in ($V_1$) one obtains a much smaller cross-sectional area in the fiber than in the preform. It should be noted that, simply because the fiber is being pulled, there will be a preferential direction induced in the fiber, as the $x$ and $y$ directions defined by this pulling cannot be equivalent. This will generally lead to a stress induced birefringence in the fiber core. This perturbation to the fiber core will be systematic, however, rather than random.

Some mention should be made of a second imperfection that occurs at the time of drawing. During deposition, there has to be an open tube for the gas flow to continue right up to the end of the process. One clearly wants to get rid of this hole during the pulling process. Indeed, the hole does disappear during the pulling process, but not generally before the heat in the interior of

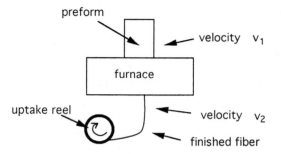

**FIGURE 5.19.** Drawing tower. A preform is fed into the furnace with linear velocity $v_1$ while a fiber from the preform material is pulled out of the oven with a linear velocity $v_2$, where the constancy of volume requires that the ratio of the preform diameter to the fiber diameter must equal the square of the ratio of $v_2$ to $v_1$.

the collapsing preform causes a burnoff of a portion of the doped layer at the interface. In the early days of fibers, this was much more of a problem than it is nowadays (the early 1990s), due to improvements in drawing technology, etc. Even when this dip was a problem, it only perturbed a small amount of energy propagating right near the fiber core.

## 5.4 COUPLED POLARIZATION IN SINGLE MODE FIBERS

We now wish to build a propagation model which takes into account both the random type perturbations we might expect from the sintering process as well as the more uniform perturbations we would expect from the drawing process. For the present, we shall analyze only single mode propagation. We assume that the transverse field in the fiber can be expressed as

$$E_t = a_x(z)f_0(r)e^{i\beta_x z}\hat{e}_x + a_y(z)f_0(r)e^{i\beta_y z}\hat{e}_y \qquad (5\text{-}50)$$

where $\beta_x \neq \beta_y$ due to birefringence. (Note the similarity to (5-49).) We further assume that $\epsilon$ is given by a tensor of the form

$$\epsilon = \begin{bmatrix} \epsilon_x & \epsilon_{xy} \\ \epsilon_{yx} & \epsilon_y \end{bmatrix} = \begin{bmatrix} \epsilon_x & 0 \\ 0 & \epsilon_y \end{bmatrix} + \epsilon_{xy}\begin{bmatrix} 0 & 1 \\ 1 & 0 \end{bmatrix} = \epsilon_f + \Delta\epsilon \qquad (5\text{-}51)$$

where the $\epsilon_f$ is the one that gives the splitting between $\beta_x$ and $\beta_y$, and that the $\Delta\epsilon$ is the random part which causes the coupling between the states and therefore gives the $z$ dependence to $a_x(z)$ and $a_y(z)$. Plugging (5-50) into

Maxwell's curl equation

$$\nabla \times \mathbf{E} = i\omega\mu_0\mathbf{H} \tag{5-52}$$

and making the slowly varying approximations

$$\left|\frac{\partial a_x}{\partial z}\right| \ll |\beta_x a_x| \tag{a}$$

$$\left|\frac{\partial a_y}{\partial z}\right| \ll |\beta_y a_y| \tag{b} \quad (5\text{-}53)$$

one finds that the transverse $H$-field $\mathbf{H}_t$ is given by

$$\mathbf{H}_t = \frac{-a_y(z)}{\eta_y} f_0(r)e^{i\beta_y z}\hat{\mathbf{e}}_x + \frac{a_x(z)}{\eta_x} f_0(r)e^{i\beta_x z}\hat{\mathbf{e}}_y \tag{5-54}$$

which just shows that the existence of modes requires the slowly varying approximation to hold.

Plugging $\mathbf{E}_t$ and $\mathbf{H}_t$ into Maxwell's curl equation

$$\Delta \times \mathbf{H} = -i\omega(\epsilon_f + \Delta\epsilon)\mathbf{E} \tag{5-55}$$

one finds that

$$\hat{\mathbf{e}}_x \frac{\partial}{\partial z}\left(\frac{a_x(z)}{\eta_x} f_0(r)e^{i\beta_x z}\right) + \hat{\mathbf{e}}_y \frac{\partial}{\partial z}\left(\frac{a_y(z)}{\eta_y} f_0(r)e^{i\beta_y z}\right)$$
$$= i\omega[\hat{\mathbf{e}}_x(\epsilon_x a_x(z)f_0(r)e^{i\beta_x z} + \epsilon_{xy}a_y(z)f_0(r)e^{i\beta_y z})$$
$$+ \hat{\mathbf{e}}_y(\epsilon_y a_y(z)f_0(r)e^{i\beta_y z} + \epsilon_{xy}a_x(z)f_0(r)e^{i\beta_x z})] \tag{5-56}$$

Making some obvious simplifications which come from the impedance relations embodied by (5-50) and (5-54), one finds the (5-56) splits into the two relations

$$\frac{\partial a_x(z)}{\partial z}\frac{f_0(r)}{\eta_x}e^{i\beta_x z} = i\omega\epsilon_{xy}a_y(z)f_0(r)e^{i\beta_y z} \tag{a}$$

$$\frac{\partial a_y(z)}{\partial z}\frac{f_0(r)}{\eta_y}e^{i\beta_y z} = i\omega\epsilon_{xy}a_x(z)f_0(r)e^{i\beta_x z} \tag{b} \quad (5\text{-}57)$$

Assuming the usual mode normalization that

$$\int \psi^2 \, d^2x = 2\eta \tag{5-58}$$

one finds the coupled mode equations

$$\frac{\partial a_x}{\partial z} = i\chi a_y e^{-i\Delta\beta z} \tag{a}$$

$$\frac{\partial a_y}{\partial z} = i\chi a_x e^{i\Delta\beta z} \tag{b}$$

$$\Delta\beta = \beta_x - \beta_y \tag{c} \tag{5-59}$$

where $\chi$ is given by

$$\chi = \int \frac{\omega\epsilon_{xy}}{2} f_0^2(r) \, d^2r \tag{5-60}$$

Note that in the normalization integral the $f_0(r)$ for the $x$ and $y$ states were assumed to be different, whereas in (5-60) they are assumed to be the same. This should be consistent with the approximations already made.

The coupled mode equations of (5-60) are really equivalent to those found in Chapter 3, Equation (3-44), for the amplitudes of the two levels of the atomic states. In that development, however, the equations contained the unknown $E(r)$ and were therefore not self-contained as is the system here. At this point we wish to point out a couple of salient features of these self-contained equations which are relevant to the problem at hand. The general solution to a system such as (5-59) is given by

$$a_x(z) = e^{-\frac{i\Delta\beta z}{2}} \left[ a_x(0) \cos bz + i\left(\frac{\Delta\beta}{2b} a_x(0) + \frac{\chi}{b} a_y(0)\right) \sin bz \right] \tag{a}$$

$$a_y(z) = e^{\frac{i\Delta\beta z}{2}} \left[ a_y(0) \cos bz - i\left(\frac{\Delta\beta}{2b} a_y(0) - \frac{\chi}{b} a_x(0)\right) \sin bz \right] \tag{b} \tag{5-61}$$

where the $b$ is given by

$$b = \sqrt{\chi^2 + \frac{\Delta\beta^2}{4}} \tag{5-62}$$

We wish to look at two limits of this equation.

Limit 1:     $\Delta\beta \gg \chi$

In this limit, Equations (5-61) yield the result that

$$a_x(z) = a_x(0) \tag{a}$$

$$a_y(z) = a_y(0) \tag{b (5-63)}$$

Therefore, if the excitation is of only one polarization eigenstate, meaning that either $a_x(0)$ or $a_y(0)$ is nonzero but not both simultaneously, then the polarization will remain constant with propagation down the fiber. Any mixture of the two states, however, will lead to propagation such as that pictured in Figure 5.16.

Limit 2:     $\chi \gg \Delta\beta$

In this limit, Equations (5-61) yield

$$a_x(z) = a_x(0) \cos \chi z + i a_y(0) \sin \chi z \tag{a}$$

$$a_y(z) = a_y(0) \cos \chi z + i a_x(0) \sin \chi z \tag{b (5-64)}$$

Clearly, if $\chi$ were a zero mean random variable with a standard deviation exceeding the value of $\Delta\beta$, the polarization state would wander from state to state all over the Poincaré sphere.

With the above discussion, we see that the additional condition necessary in order to have stable polarization during propagation is that the fiber have high birefringence. This is achievable in practice, but not so easily. One way to achieve high birefringence is depicted in Figure 5.20. The idea here is to

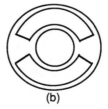

(a)                                    (b)

**FIGURE 5.20.**    Differential doping profile in (a) single and (b) multimode bowtie fibers, illustrating why the technique can be applied to single but not to multimode fibers.

dope the cladding differentially so that a very different stress distribution is induced in the $y$ direction with respect to the $x$ direction. Due to the appearance of the doping distribution, these fibers are often referred to as *bowtie fibers.* The point of having (a) and (b) parts to the figure is to illustrate that high birefringence single mode fibers are feasible to make, but high birefringence multimode fibers are not. In the single mode case, the core diameter may be less than 10 $\mu$m, while the outer diameter is circa 125 $\mu$m. This leaves a lot of space for differential doping to try to achieve a reasonably uniform stress distribution within the core. In the multimode case, the core diameter may be 62.5 $\mu$m compared to the outer diameter of 125 $\mu$m. This makes it hard enough to achieve a uniform stress distribution in the core period, let alone one that would induce the same $\beta_x\beta_y$ propagation constant splitting for all modes. As was discussed above, there will be birefringence in the multimode fiber such that the modes can perhaps best be expressed as LP modes. However, the $\chi$'s and $\Delta\beta$'s for each mode are likely to be quite different, and therefore the polarization is quite likely to be quite random after as little as a few meters of propagation. Some consequences of this will be taken up in the next section.

## 5.5  PROPAGATION IN FIBERS IN GENERAL

This section will concentrate mostly on propagation in multimode fibers, as the most important characteristics of single mode fiber propagation are polarization evolution, just discussed, and dispersion, to be discussed in the next section. Multimode propagation is an interesting topic, and also instructive in that it gives us a wealth of information about how large aggregates of modes interact together. But before we jump into the topics of modal noise, mode continuum propagation, and fiber coupling, we first wish to discuss in detail some salient characteristics of fibers.

As we first noticed in Equation (5-40), the modes of a multimode parabolic index fiber may be (approximately) expressed in the form

$$\beta_{\mu\nu} = k_1 \left[ 1 - 2\frac{2\Delta}{V}(2\mu + \nu + 1) \right]^{1/2} \tag{5-65}$$

Later in this section, we shall see that a comparable expression can be derived for any $\alpha$ index profile. We know that the propagation constant $\beta_{\mu\nu}$ for bound modes must be bounded between the core propagation constant $k_1$ and the cladding propagation constant $k_2$, which is given by

$$k_2 = k_1[1 - 2\Delta]^{1/2} \tag{5-66}$$

Combining (5-65) and (5-66), we notice that the cutoff condition is express-ible in the alternative form

$$m = 2\mu + \nu + 1 \le \frac{V}{2} \tag{5-67}$$

where $m$ is often referred to as the *principal mode number*. Figure 5.21 gives some more meaning to $m$. From (5-65), all modes with a given mode number $m$ have the same propagation constant and therefore belong to a mode group. Looking at the left of the $\mu$ axis and below the $\nu$ axis, we see that the $m^{th}$ mode group contains $2m$ total modes. If cutoff occurs at some $m = m_{MAX} = \left[\dfrac{V}{2}\right]$, the total number of modes $N$ bound by the fiber must be given by

$$N = 2 \sum_{m=1}^{m_{MAX}} M = m_{MAX}[m_{MAX} + 1] \tag{5-68}$$

which, for a large enough $m_{MAX}$, can be approximated as

$$N \sim \frac{V^2}{4} \tag{5-69}$$

Equation (5-68) would predict that, for single mode (actually double mode

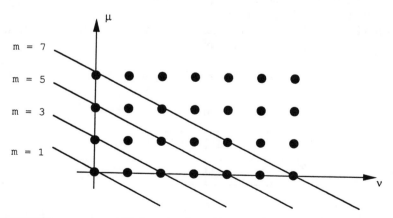

**FIGURE 5.21.** $\mu$-$\nu$ space with the $\mu\nu$ mode positions drawn in as dots and the constant $m$ contours as lines, and where each $m$ contour is labeled as to $m$ number as well as the total number of modes for the $m$ number.

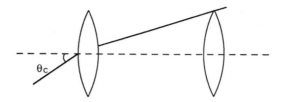

**FIGURE 5.22.**    Meaning of the concept of numerical aperture NA for a lens system for which the NA $= \sin \theta_c$, where $\theta_c$ is the minimum input angle for which a centrally located ray will fail to pass through the system.

due to polarization) propagation, one would need a $V \le 2.0$. If one were to use the dispersion relation for the step index hybrid modes, one would find that the $V$ should really be less than about 2.4.

Before considering some examples of typical $a$'s and $\Delta$'s for fibers, it is necessary to give some more meaning to the $\Delta$ of (5-65). Figure 5.22 reviews the meaning of the numerical aperture for a free-space lensing system as was discussed in the second chapter. In a step index fiber, the critical angle $\theta_c$ is defined analogously to the $\theta_c$ of Figure 5.22, as illustrated in Figure 5.23. We now wish to find the fiber's numerical aperture, NA, in terms of the index contrast $n_1 - n_2$ and thereby relate NA and $\Delta$. The reason why this is desirable is illustrated in Figure 5.24. If one simply excites a fiber with a source of much larger NA than the fiber's NA, then the fiber will guide all modes with a radiation angle up to its NA, and therefore the fiber NA is readily measured by simply measuring the maximum radiation angle at the fiber output. For this reason, fibers are normally supplied with data sheets specifying their diameter and numerical aperture.

From Figure 5.23, one readily sees that, from writing the Fresnel relation at the fiber input, the NA is given by

$$NA = n_1 \sin \theta_1 \qquad (5\text{-}70)$$

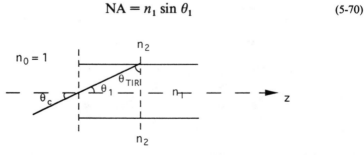

**FIGURE 5.23.**    Definition of the critical excitation angle for a step index fiber whose NA will be defined as $\sin \theta_c$, where $\theta_c$ is the input angle for which a ray incident on the fiber at the fiber center at ray angle $\theta = \theta_c$ will be incident on the core cladding interface at a normal angle equal to the total internal reflection angle.

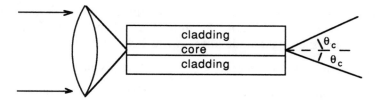

**FIGURE 5.24.** Operational definition of the fiber NA whereby NA is simply given as the sin of the maximum radiation angle of fiber with an "overfilled" excitation.

From the total internal reflection condition for an $n_1 - n_2$ interface, one can write that

$$n_1 \sin \theta_{TIR} = n_2 \qquad (5\text{-}71)$$

From the right triangle with angles $\theta_{TIR}$ and $\theta_1$, one can write that

$$\sin^2 \theta_i = 1 - \sin^2 \theta_{TIR} \qquad (5\text{-}72)$$

and thereby obtain that

$$NA = \sqrt{n_1^2 - n_2^2} \qquad (5\text{-}73)$$

From the relation that

$$n_2^2 = n_1^2(1 - 2\Delta) \qquad (5\text{-}74)$$

one can see immediately that

$$NA = n_1 \sqrt{2\Delta} \qquad (5\text{-}75)$$

We are now ready to compute mode numbers and fiber radii from data sheets. Say, for example, that we had a data sheet for an early 1980s telecom fiber of 50 $\mu$m diameter and NA = 0.2. Assuming an operation wavelength of 0.85 $\mu$m (first window), one could use Equation (5-69) to determine that the number of modes $N$ bound in this fiber would be roughly 300. As most of the multimode fibers currently in use are of larger core diameters and NA's than these early ones, we can consider this $N$ to be a rough lower bound for multimode fibers. We can also play the game the other way. It is much harder to measure core diameter for a single mode fiber, as the single mode nature dictates that the spot must be diffraction limited. Let us say, therefore, that a numerical aperture measurement of a fiber performed at 1.3 $\mu$m indicated that the fiber was indeed single mode and with an NA of roughly 0.1.

Recalling the argument which followed Equation (5-69), which stated that single mode cutoff occurred for $V$ numbers less than roughly 2.4, we may conclude that the single mode fiber on which we performed the near field measurement has a diameter of $a < \dfrac{2.4}{NA\, k_1}$, which gives a value of $a < 5 \ \mu m$.

A very important characteristic of propagation in a fiber is that the energy decays as it propagates down the fiber. This loss comes from a variety of causes. For example, in a system one has to cable a fiber as well as splice and connectorize it. Cables cause bends which in turn cause bending loss due to the fact that the fiber is no longer symmetric about the $z$ axis. Splices and connectors by nature are imperfect and therefore also couple energy out of the core. For the moment, however, for simplicity let us concentrate attention on losses intrinsic to a straight, bare fiber. Here the loss will be due mainly to absorption and scattering. A typical plot of loss versus wavelength is given in Figure 5.25, which contains a wealth of information about the causes of the losses. Perhaps a few words should be said about the measurement of the curve before making too much of it, however. Both scattering and absorption are going to be dependent on impurities in the glass where dopants must be included as an impurity type. This means that the loss should, by nature, vary across the core. Further, the change in composition of the glass at or near the core-cladding interface, including any surface roughness, should lead to increased scattering from this interface. Fortunately, the effect of radially varying dopants is not nearly as pronounced as that of core-cladding interface roughness or, for that matter, scattering from the dip. Correcting for all of these mode dependent attenuation characteristics has been one of the big problems in multimode fiber attenuation measurement. However, if we ignore scattering from the dip as involving only a small portion of the power and scattering variation due to doping as a small perturbation, then we need worry only about core-cladding interface scatter-

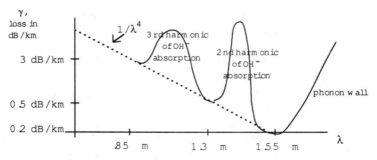

**FIGURE 5.25.**    Typical plot of loss versus wavelength for a low loss fused silica fiber.

ing. We completely ignore absorption effects in this argument, as we would not want to propagate in a spectral regime where absorption is prevalent. Figure 5.26 helps to illustrate what happens in this case. The idea is that, if there is a big difference in attenuation coefficient, the modes with the much greater attenuation coefficient simply disappear at some distance. Therefore one can use the concept of steady-state distribution to try to define a mode independent attenuation coefficient. If one were using such a distribution, therefore, one could write that the power $P_{\mu\nu}(z)$ in the $\mu\nu^{\text{th}}$ mode at the longitudinal coordinate $z$ would be given by

$$P_{\mu\nu}(z) = P_{\mu\nu}(0)e^{-\gamma z} \tag{5-76}$$

where $\gamma$ is the mode independent attenuation coefficient, which was plotted versus $\lambda$ in Figure 5.25.

Returning to the discussion of Figure 5.25, we see several noticeable features on this sketch. The $1/\lambda^4$ curve upon which the whole curve is riding is reminiscent of the Rayleigh scattering coefficient which causes the sky to appear blue on sunny days. Indeed, as the atmosphere is made up of particles much smaller than an optical wavelength, so are the density fluctuations of the amorphous material comprising the glass of the fiber. Therefore, in the absence of absorption, the loss is due almost wholly to Rayleigh scattering. There are two pronounced absorption peaks on the plot, which correspond to the second and third harmonics of the hydroxyl ion resonance at 2.85 $\mu$m. Due to the structure of $SiO_2$, it is essentially impossible to reduce the hydroxyl content to zero, as water can nestle anywhere into the silicon dioxide matrix. As it would be ridiculously expensive to have to draw fibers under totally anaerobic conditions, one learns to live with the peaks and the three propagation windows they define, one about 0.85 $\mu$m, which corresponds closely with the bandgap of GaAs; one about 1.3 $\mu$m, which corresponds closely to the dispersion minimum of fused silica; and one at 1.55 $\mu$m, which represents the loss minima of fused silica as phonon absorption becomes prevalent for wavelengths above 1.6 $\mu$m.

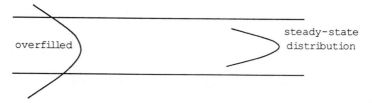

**FIGURE 5.26.** Evolution of the spatial distribution of energy in a multimode over the first length of propagation from the input end.

We now seem ready to consider problems in fiber propagation. The transverse field inside a multimode optical fiber excited by a monochromatic (and therefore coherent) source can be expressed in the form

$$\mathbf{E}_t(r) = \sum_{\mu\nu} \psi_{\mu\nu}(r)[(A^c_{x\mu\nu}\cos\nu\theta + A^s_{x\mu\nu}\sin\nu\theta)\hat{\mathbf{e}}_x e^{i\beta_{x\mu\nu}z}$$
$$+ (A^c_{y\mu\nu}\cos\nu\theta + A^s_{y\mu\nu}\sin\nu\theta)\hat{\mathbf{e}}_y e^{i\beta_{y\mu\nu}z}] \quad (5.77)$$

where it has been assumed that there is little splitting between orientational modes yet enough splitting between polarization modes that the LP modes are the actual modes. The $A^c_{x\mu\nu}, A^s_{x\mu\nu}, A^c_{y\mu\nu}, A^s_{y\mu\nu}$ are determined by the initial conditions, as will be further discussed later in this section. As we saw in the development leading to Equations (5-59), small perturbations in fiber geometry, etc., lead to equations which couple the amplitudes of different modes. Equations (5-59) are parameterized by two parameters, $\chi$ and $\Delta\beta$. As was seen in the discussions following Equation (5-61), if the relative splitting between modes, $\Delta\beta$, greatly exceeds the perturbation induced coupling coefficient, $\chi$, then mode coupling does not occur. This is not to say that propagation becomes length independent, as modal beating still occurs. In the opposite limit, where the modes are nearly degenerate with respect to the magnitude of the coupling coefficient, the coupling is free and most likely random. Here, we will make an assumption as to the size of $\chi$ relative to the mode spacings of, for example, Figure 5.13 and stick to it throughout what follows. The basic assumption is that $\chi$ is less than the spacing between adjacent mode groups. This was not the case in the early days of low-loss fibers (early 1970s) but became true in the late 1970s, and the $\chi$'s intrinsic to the fiber naturally seem to reduce with time and improved technology. Now, for $\chi$ less than this splitting, there is little or no mode coupling between adjacent mode groups. As it has been assumed that the orientational states are degenerate, these will couple randomly. In actuality, these orientational states probably are split a little, although maybe not so much as the polarization states. However, for the modes to become LP rather than hybrid, this splitting may be very small, as the "strongly-guiding" splitting would not even show up on Figure 5.13. We can therefore assume that the modes are LP and still assume that the polarization splittings are less than $\chi$. This we will do. The one sticky point left is the question as to whether there is coupling within different $\mu\nu$ states of a mode group. As we shall see in subsequent paragraphs, the question is moot, as it makes no difference to modal noise statistics or mode continua arguments. We therefore pick a position and stick to it. As indicated in Figure 5.13, the imposition of a boundary condition on the hybrid modes causes a splitting of $\mathbf{HE}_{12}$ from $\mathbf{EH}_{11}$. It therefore seems safe to assume that the imposition of a boundary condition on the LP solutions (no infinite profile assumption) would split the $LP_{10}$ from the $LP_{02}$. We further assume that this splitting is sufficiently large that mode coupling is greatly inhibited between these members of a degenerated mode group. There is experimental evidence to indicate that this may be the case. Following the above discussion, we can begin to make some statements about the power flow in the fiber. Assuming that we are considering a regime of quasi-steady state, then Equation (5-76) must apply. Recalling that

$$P_{\mu\nu}(0) = \frac{1}{\eta_{\mu\nu}} \int |E_{t\mu\nu}|^2 \, d^2x \qquad (5-78)$$

one can write (5-76) in the form

$$|A^c_{x\mu\nu}|^2 + |A^s_{x\mu\nu}|^2 + |A^c_{y\mu\nu}|^2 + |A^s_{y\mu\nu}|^2 = P_{\mu\nu}(0)e^{-\gamma z} \qquad (5-79)$$

where mode orthogonality has been assumed and $\gamma$ is explicitly independent of $\mu$ and $\nu$. Therefore, one can conclude that the $A$'s will be randomly distributed but must satisfy the constraint of (5-78).

As the $A$'s are randomly distributed, we need some kind of statistical model to describe the propagation. The propagation problem as described by Equation (5-77) appears to be deterministic, at least as far as the phrases are concerned, but this is not really true, as the $A$'s contain randomly varying phases. It should be pointed out here, however, that the randomness of the $A$'s is with distance and not time. When we think about this a little, however, we might begin to become convinced that the whole phase problem is a stochastic one. For example, if we have 300 modes, to calculate the intensity will require beating between all of these modes, or $300^2/2 = 45,000$ terms. Calculation of each one of these terms will require knowledge of the propagation distance $z$, which can be on the order of kilometers, to wavelength accuracy. Even if this could be achieved, the differences in the propagation constants would have to be known to comparable accuracies ($\mu$m/km $= 10^{-11}$) for the phases to track. On top of these problems is the problem that the coupling of the $A$'s in a $\mu\nu$ mode cause phase perturbations. It is very unlikely that a deterministic prediction of the interference pattern of a fiber could be especially accurate after any propagation distance greater than a couple of beat lengths, and then only if one could accurately predict the distribution coupled into the fiber at the input.

We can use the above considerations to calculate the field and intensity statistics at a point on the fiber endface. In a given polarization state, given the above considerations, one can express the field in the form

$$E_n(x,y) = \sum_{j=1}^{N} a_j \psi_j(x,y) e^{i\phi_j}, \qquad n = x,y \qquad (5\text{-}80)$$

where the resultant point intensity could be expressed as a sum over the intensities in the two polarization states. As Equation (5-80) is in the form of the random walk sum, we can make statements about the statistics of the problem. We consider the ensemble to be averaged over to be a set of different $z$-values — that is, effective fiber lengths. Further, we assume that there are enough modes that the law of large numbers applies. For the $\phi_j$'s uniformly distributed, it is clear that the average value of the field will be

$$\langle E_n(x,y) \rangle = 0 \qquad (5\text{-}81)$$

Clearly, the average intensity value will therefore be given by

$$I(x,y) = \sum_{j=1}^{N} a_j^2 \psi_j^2(x,y) \qquad (5\text{-}82)$$

which, from the law of large numbers, will be Poisson distributed for a single polarization state, as illustrated in Figure 5.27. The distribution will be gamma if both polarization states are sensed (Hjelme and Mickelson 1983). For the field to have a zero mean at any given point means that the field must change sign many times as the length is varied. As different modes have different amplitudes across the core, the admixture of random phases at different points on the core is also different. The situation is formally similar to that encountered in standard laser speckle, as illustrated in Figure 5.28. It is easily seen that the field at a point $x,y$ in Figure 5.28 must be given by the

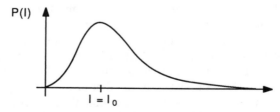

**FIGURE 5.27.**    Probability distribution for the intensity measured at a point on the end-face of a multimode fiber single polarization excited by a coherent monochromatic source.

reflective surface

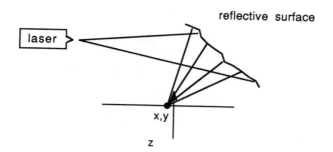

laser

x,y

z

**FIGURE 5.28.**    Multiple and random contributions to the field on the observation plane due to the scatter of laser light from a rough, partially reflective surface.

sum

$$E_n(x,y) = \sum_j a_j e^{i\phi_j} \qquad (5\text{-}83)$$

where the $a_j$'s and $\phi_j$'s are random. As is clearly evident there, motion along either the $z$ axis or in the $x,y$ plane will cause a random field change and therefore will lead to changes in sign of the field, with the attendant nulls between the maxima. The decorrelation lengths in the $z$ and $x$-$y$ directions will in general be different, as they are geometry dependent, but the effect is qualitatively the same. Due to the changes in sign of the field, a one-dimensional scan of the intensity might look like Figure 5.29. The pattern seen on the output end of a circular optical fiber therefore might look like Figure 5.30. The pattern at different planes will, in general, be different, so that an average over an ensemble of $z$'s wil give one back the average value of the intensity. It may further be pointed out without derivation that the average number of speckles across the output facet of a fiber is equal to the number of "equally" excited modes in the fiber. So, what we have seen is that, if we excite a multimode fiber with a monochromatic laser, we get a noiselike output. However, this pattern is stationary in time. Further, if we were to

I(x)

**FIGURE 5.29.**    Possible intensity scan through a speckle pattern.

**FIGURE 5.30.**    Possible realization of a speckle pattern at the output of an optical fiber.

measure the total power carried by the fiber by focusing the whole fiber output into a detector, cut the fiber back by a length such that $\alpha z$ is small (i.e., cutback of less than a few meters), then repeat the same measurement, Equation (5-79) tells us that we would measure the same total power carried by the fiber. However, if in these measurements we had put an aperture across the detector face which eclipsed a portion of the fiber core, we would not in general have obtained the same result, as we would have subtended a different number of speckles. In general, any core position dependent stop or so-called mode dependent filtering would have had the same effect, that of changing the effective power transmitted due to the different number of speckles being vignetted.

Unfortunately, the speckle pattern is really not always so stationary as the above considerations might lead us to think. Consider a slab of glass of length $L$ in thermal equilibrium, as in Figure 5.31. The optical path length $L_{opt}$, which is given by the index $n$ multiplied by the length $L$, will be a function of temperature. Indeed, a change in temperature $\Delta T$ will induce an optical path length change $\Delta L_{opt}$ given by

$$\frac{\Delta L_{opt}}{L_{opt}} = \left( \frac{1}{L}\frac{dL}{dT} + \frac{1}{n}\frac{dn}{dT} \right) \Delta T \qquad (5\text{-}84)$$

For fused silica glass, the characteristic $\dfrac{1}{L}\dfrac{dL}{dT}$ is roughly $10^{-6}/°C$, while $\dfrac{1}{n}\dfrac{dn}{dT}$

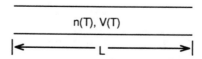

n(T), V(T)

$\mid\!\longleftarrow\!\quad L \quad\!\longrightarrow\!\mid$

**FIGURE 5.31.**    Slab of glass of length $L$, index $n(T)$ and volume $V(T)$ in equilibrium at a temperature $T$.

is roughly $-10^{-5}/{}^\circ$C. Using only the $dn/dT$ value, one can show that

$$\frac{\Delta L_{opt}}{\lambda} = 10^{-5}\left(\frac{L_{opt}}{\lambda}\right)\Delta T \tag{5-85}$$

Using values like $L_{opt} \sim 1$ km and $\lambda \sim 1$ $\mu$m yields the expression that

$$\frac{\Delta L_{opt}}{\lambda} = 10^4 \Delta T \tag{5-86}$$

showing that a change of temperature of $10^{-4}$ °C can cause wavelength path changes in the fiber. Perhaps a more telling number would be the change in temperature required to change the phase between the lowest and the highest order mode by $2\pi$. The required length change would be

$$\Delta L = \frac{2\pi}{k_2 - k_1} \tag{5-87}$$

and would correspond to a temperature change of $10^{-2}$ °C. Such changes can be avoided under laboratory conditions, at least in the short term, but not under field conditions. In fact, it is just such effects which make the optical fiber such a good choice as a sensing element, especially in interferometric sensor configurations. These index changes effectively change the fiber length and therefore give a time-varying speckle pattern in the present case. As was mentioned above, any component which performs a mode dependent filtering operation will therefore have a time-varying transmission. This effect of time-varying transmission is generally known as *modal noise* and can be disastrous to information transfer over fiber cables. It should further be pointed out that temperature fluctuations are not the only cause of modal noise, which can also be caused by, for example, frequency variations. This frequency chirp effect has actually been applied to the measurement of fiber dispersion.

If one idea comes out of the above discussion, it should be that one does not want to use an "overly" monochromatic source with a multimode fiber. The question therefore arises as to how monochromatic is monochromatic, as well as how to analyze a fiber excited by an incoherent source. Typical spectra for three GaAlAs sources operating in the short wavelength window are given in Figure 5.32. We assume that the single mode source can be considered monochromatic. This assumption will later be demonstrated to be acceptable. However, multimode sources, luminescent sources, or light emitting diodes (LED's) are a different case. Generally what we do to solve

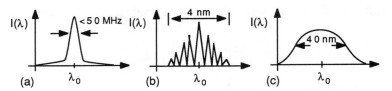

**FIGURE 5.32.** Mode structures for three different types of GaAIAs laser sources: (a) a single mode laser, (b) a multimode laser, and (c) a light-emitting diode (LED).

Maxwell's equations is to consider the solution for a monochromatic excitation by taking that

$$\mathbf{E}(\mathbf{r},t) = \text{Re}[\mathbf{E}_\omega(\mathbf{r})e^{-i\omega t}] \qquad (5\text{-}88)$$

and solving for $\mathbf{E}_\omega(\mathbf{r})$. Then given a complex source amplitude spectrum $A(\omega)$, one can form the time variation of the total electric field $\mathbf{E}_{tot}(\mathbf{r},t)$ by

$$\mathbf{E}_{tot}(\mathbf{r},t) = \int \mathbf{E}_\omega(\mathbf{r})A(\omega)e^{-i\omega t}\,d\omega \qquad (5\text{-}89)$$

The problem with this expression is, of course, that we can't possibly know the phases of the excitation spectra for either a multimode source or an LED. Therefore our approach to the modal problem is in real trouble. In the next paragraph we shall try to find a way around this problem.

Consider the geometrical optics of an optical fiber for the present in the hope that a bridge can be found back to the modal propagation picture. To remind the reader, in geometrical optics what one wishes to do is to find the coordinates $\mathbf{r}$ of a ray path where this ray path represents the evolution in space of the normal to a phase front of a local plane wave travelling through a slowly varying (compared to a wavelength) medium. The situation is as

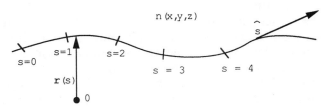

**FIGURE 5.33.** Ray path in a spatially varying medium of index $n(x,y,z)$, where $s$ is the distance measured along the ray path $\hat{s}(s)$ is the unit vector pointing in the direction of the ray path at coordinate $s$ and $\mathbf{r}(s)$ is the vector describing the ray path.

depicted in Figure 5.33, where $s$ is the distance measured along the ray path, defined as

$$s = \int_0^s ds = \int_0^s \sqrt{(dx)^2 + (dy)^2 + (dx)^2} \qquad (5\text{-}90)$$

and the unit vector $\hat{s}$ along the ray path is defined as

$$\hat{s} = \frac{d\mathbf{r}}{ds} \qquad (5\text{-}91)$$

and the equation for the coordinate of the ray $\mathbf{r}(s)$ is given by

$$\frac{\partial}{\partial s} [n(\mathbf{r}) \frac{d\mathbf{r}}{ds}] = \nabla n(\mathbf{r}) \qquad (5\text{-}92)$$

The ray path in an optical waveguide should be nothing more than the effective normalized $k$ vector of a local plane wave in that medium. As we know that fibers are weakly guiding, we know that modes are plane-wavelike, and a ray representation should not be too far from reality. Writing our radius vector as

$$\mathbf{r} = r\hat{\mathbf{e}}_r + z\hat{\mathbf{e}}_z \qquad (5\text{-}93)$$

we find that

$$\hat{s} = \frac{\mathbf{k}}{|\mathbf{k}|} = \frac{dr}{ds} \hat{\mathbf{e}}_r + r \frac{d\phi}{ds} \hat{\mathbf{e}}_\phi + \frac{dz}{ds} \hat{\mathbf{e}}_z \qquad (5\text{-}94)$$

where the situation is as depicted in Figure 5.34, which defines the components of the $k$ vector. Note that the $z$ component of the $k$ vector is just $\beta$, the propagation constant, and therefore the $z$ component of (5-92) should give us some bridge to wave optics. From (5-94), we can immediately obtain

$$k_r = n(r)k_0 \frac{dr}{ds} \qquad \text{(a)}$$

$$k_\phi = n(r)k_0 n \frac{d\phi}{ds} \qquad \text{(b)}$$

$$k_z = n(r)k_0 \frac{dz}{ds} = \beta \qquad \text{(c)} \quad (5\text{-}95)$$

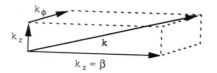

**FIGURE 5.34.**    Components of the $k$ vector.

Taking our index to be an $\alpha$ profile, as per Equation (5-24), one can write out (5-92) in the form

$$\frac{d}{ds}\left(n(r)\frac{dr}{ds}\right) - rn(r)\left(\frac{d\phi}{ds}\right)^2 = \frac{dn(r)}{dr} \qquad \text{(a)}$$

$$\frac{d}{ds}\left(r^2n(r)\frac{d\phi}{ds}\right) = 0 \qquad \text{(b)}$$

$$\frac{d}{ds}\left(n(r)\frac{dz}{ds}\right) = 0 \qquad \text{(c)} \quad (5\text{-}96)$$

To solve Equations (5-96), we begin with the last equation first. We note that this equation can be rewritten in the form

$$n(r)k_0\frac{dz}{ds} = \text{constant} = \beta \qquad (5\text{-}97)$$

where the second equation in (5-97) comes from Equation (5-95c). Equation (5-97) is indeed the one which should form the bridge between ray optics and geometrical optics. To see how, we first note that $dz/ds$ must be the cosine of an angle, that angle representing the angular direction of a ray with respect to the $z$ axis. Using this fact, together with Equation (5-24), one can express (5-97) in the form

$$\frac{\beta^2}{k_1^2} = \left[1 - 2\Delta\left(\frac{r}{a}\right)^\alpha\right][1 - \sin^2\theta(z)] \qquad (5\text{-}98)$$

Weakly guiding implies both that $\Delta$ is small and that $\theta^2(z)$ is of the same order. Using this fact, one can expand (5-98) to obtain the form

$$R^2 = \frac{1}{2\Delta}\left[1 - \frac{\beta^2}{k_1^2}\right] = \left(\frac{r(z)}{a}\right)^\alpha + \frac{\sin^2\theta(z)}{2\Delta} \qquad (5\text{-}99)$$

where $R$ is the mode parameter which is indeed the bridge between ray and wave optics. Clearly, the term in $\beta$ on the right-hand side of the $R^2$ expression is a constant of propagation. This means that the right-hand side of the last second equation in (5-99) defines an invariant along a ray path. As sketched in Figure 5.35, a constant $R$ parameter defines rays whose angles must decrease with increasing radius from the axis. This is by definition a bound mode. Further, $R$ is clearly bounded between 0 (lowest order mode) and 1 (cutoff).

Returning to Equation (5-96), clearly we may rewrite it in the form

$$r^2 k_0 n \frac{d\phi}{ds} = \text{constant} = v \tag{5-100}$$

where $v^2$ is simply what we choose to name the constant, with the use of some hindsight (i.e., Equations (5-30) and (5-31)). But (5-100) clearly is equivalent to the statement that

$$k_\phi = \frac{v}{r} \tag{5-101}$$

Recalling that

$$k_0^2 n^2(r) = k_r^2 + k_\phi^2 + \beta^2 \tag{5-102}$$

one can show that

$$k_r^2 = k_0^2 n^2(r) - k_\phi^2 - \beta^2 \tag{5-103}$$

which obviates the need to solve Equation (5-96a) for $k_r$. Clearly, Equation (5-103) is equivalent to

$$n(r) k_0 \frac{dr}{ds} = \sqrt{k_0^2 n^2(r) - \frac{v^2}{r^2} - \beta^2} \tag{5-104}$$

**FIGURE 5.35.**    Axial ray path in an $\alpha$ profile index fiber.

which, with the aid of the expression

$$\frac{d}{ds} = \frac{dz}{ds}\frac{d}{dz} = \frac{\beta}{nk_0}\frac{d}{dz} \tag{5-105}$$

reduces to the quadrature

$$\int dz = \int \frac{\beta\, dr}{\sqrt{k_0^2 n^2(r) - \dfrac{v^2}{r^2} - \beta^2}} \tag{5-106}$$

which can be used to solve for the ray paths.

Our purpose at this point is not to find exact solutions for ray paths but to try to find a technique to handle finite linewidth sources, as such sources seem to be the only ones we can practically use to excite multimode fibers. Perhaps a way to get at this problem is to use our bridge to wave optics and see qualitatively what happens as linewidth increases. The greatest effect of wave optics is that it only allows certain values of $\beta$, the propagation constant, to exist inside the waveguide. Together with Equation (5-99), this implies that only certain $R$ values, and therefore only certain ray paths, can exist inside the waveguide. If one were to Fourier analyze the $\beta$ spectrum of the field inside of a multimode fiber, with parabolic square line index profile, one would find delta spikes at a spectrum of $\beta$ values, which would be spaced by approximately

$$\Delta\beta = \beta_{m+1} - \beta_m = n_1 k_0 \sqrt{1 - 2\Delta\left(\frac{m+1}{\sqrt{N}}\right)} - n_1 k_0 \sqrt{1 - 2\Delta\left(\frac{m}{\sqrt{N}}\right)}$$

$$\approx -\frac{\sqrt{2\Delta}}{a} \tag{5-107}$$

where $N$ is the total number of bound modes as is given by $\sqrt{2\Delta}n_1 k_a$. However, the positions of the $\beta$ values are directly affected by the wavelength of the source through the factor $k_0$. In fact, a source of linewidth $\delta\lambda$ at a center frequency $\lambda$ would induce a $\beta$ space spread of $\delta\beta$, which is given by

$$\delta\beta = \beta(\lambda) - \beta(\lambda + \delta\lambda) = \bar{n}_1 k_0 \frac{\delta\lambda}{\lambda} \tag{5-108}$$

where $\bar{n}_1$ is the group index. The situation is depicted in Figure 5.36. Clearly, if $\delta\beta \gg \Delta\beta$, then the $\beta$ spectrum will become a continuum between the

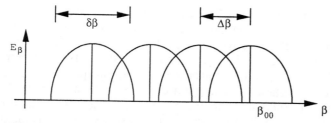

**FIGURE 5.36.**    $\beta$ spectrum of the multimode field in a fiber for monochromatic and finite width excitation.

cutoff points, as depicted in Figure 5.37. If this is the case, then Equation (5-99) no longer predicts discrete $R$ values, but all $R$ values are allowed. This implies that a continuum of rays is allowed and therefore that, indeed, ray optics is the correct theory to apply in this mode continuum limit.

It is easy to show that, if $\delta\lambda$ satisfies

$$\frac{\delta\lambda}{\lambda} > \frac{\sqrt{2\Delta}}{\bar{n}_1 k_0 a} = \frac{2\Delta}{V} = \frac{1}{n_1^2} \frac{NA^2}{V} \tag{5-109}$$

then the mode continuum applies. Using the typical numbers for a 50 $\mu$m diameter, 0.2 NA fiber of $V = 35$, one finds that the limit on $\delta\lambda$ is

$$\frac{\delta\lambda}{\lambda} > 5 \times 10^{-4} \tag{5-110}$$

This will clearly not be satisfied for the single mode laser of Figure 5.32(a), where $\delta\lambda/\lambda \sim 10^{-7}$. However, for either the multimode laser of that figure ($\delta\lambda/\lambda \sim 5 \times 10^{-3}$) or the LED ($\delta\lambda/\lambda \sim 5 \times 10^{2}$), the mode continuum limit applies, and one can safely use geometrical optics as a "correct" theory.

Before turning the discussion to dispersion, we now discuss some practical

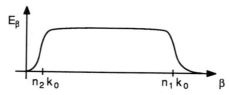

**FIGURE 5.37.**    $\beta$ spectrum for a waveguide excited with a source whose spectral width induces a $\beta$ spread $\delta\beta$ which greatly exceeds the $\beta$ spaces $\Delta\beta$.

effects of source coherence. We have already seen that source coherence can have a disastrous effect on multimode propagation. In the mode continuum limit, however, the effective number of modes becomes infinite, and therefore the relative speckle size (i.e., the speckle visibility) should approach zero and modal noise should no longer be an effect. The question then arises as to what kind of excitation one might use with a single mode fiber. We take up that issue in the following paragraph. Let us consider the excitation problem depicted in Figure 5.38. Here we consider disturbance $E_{inc}(r)$ incident from free space onto a fiber endface. For simplicity we assume the cladding, of index $n_2 = n_1 \sqrt{1 - 2\Delta}$, to be infinite in extent, and we only consider the weakly guiding limit. Let us first consider the monochromatic excitation case. In this case, the expression for the bound electric field $E_b$ is given by

$$E_b(r) = \left[ \sum_{\mu\nu} a_{\mu\nu}\psi_{\mu\nu}(x,y,z)\hat{e}_x + \sum_{\mu\nu} b_{\mu\nu}\psi_{\mu\nu}(x,y,z)\hat{e}_y \right] e^{i\beta_{\mu\nu}z} \qquad (5\text{-}111)$$

where the $\psi_{\mu\nu}(x,y,z)$ are defined in Equation (5-43). It is our present purpose to find the $a_{\mu\nu}$'s and $b_{\mu\nu}$'s, as we really do not need the details of the radiated modes, which are lumped into the $E_{rad}$ term.

To solve such a boundary value problem in all its generality is not possible. The radiation is an uncountably infinite sum of modes. Therefore, on the boundary $z = 0$, we have two equations in one known (the amplitude of $E_i$) and an infinite number of unknowns. By playing with mode orthogonality, we could conceivably turn this system into an uncountably infinite number of equations (an integral equation, perhaps) for the uncountably infinite number of unknowns. Theoretically, such a system is solvable, but clearly not in this course. Instead, we shall use the weakly guiding approximation to find the simplified overlap integral picture of excitation.

Weak guidance means that, to zeroth order, the interface of Figure 5.38 is a plane interface between two media of indices $n_0 = 1$ and $n_2$. Further, weak guidance precludes high angle rays from guiding, and therefore any plane waves propagating at angles significantly higher than the normal to the

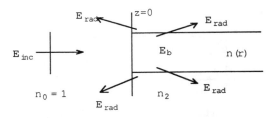

**FIGURE 5.38.**    The fiber excitation problem.

interface will couple only to radiation modes and be of no interest to us here. As any excitation can be decomposed into plane waves, we need only consider "nearly" normal excitations. Therefore, the zeroth order solution to the excitation problem is that the incident wave's amplitude is effectively reduced by a factor $1 - \left(\dfrac{n_2 - n_0}{n_2 + n_0}\right)$ due to the Fresnel reflection. This reduced wave is the one just to the right of the interface. The $E_b$ and $E_{rad}$ just to the right of this interface should form a complete set, and therefore we can decompose the reduced incident wave into this basis. The problem is formally given by

$$\frac{2n_0}{n_2 + n_0}\, \mathbf{E}_{inc}(x,y,z = 0^+) = \mathbf{E}_b(x,y,z = 0^+) + \mathbf{E}_{rad}(x,y,z = 0^+) \quad (5\text{-}112)$$

Before solving (5-112) for the $a_{\mu\nu}$'s and $b_{\mu\nu}$'s, we can make two simplifications. One is to ignore the $\dfrac{2n_0}{n_2 + n_0}$ term, as we are interested in relative values of the $a_{\mu\nu}$'s and $b_{\mu\nu}$'s here. If we were interested in absolute values we would need to include it, but then again, if we were interested in absolute values we would probably anti-reflection coat the fiber endface. Secondly, we are not interested in the radiation here. The only effect of ignoring the $\mathbf{E}_{rad}$ in (5-112) will be that we'll lose track of a portion of the energy that was supplied by $\mathbf{E}_{inc}$. No error will be incurred, however, in the $a_{\mu\nu}$'s or $b_{\mu\nu}$'s. With these simplifications, one can solve Equation (5-112) by first multiplying both sides by $\psi_{\mu\nu}(x,y,z = 0^+)\hat{e}_x$ and then multiplying both sides by $\psi_{\mu\nu}(x,y,z = 0^+)\hat{e}_y$ and integrating. The results are

$$a_{\mu\nu} = \frac{1}{2\eta_{\mu\nu}} \int_{x\ section} \mathbf{E}_{inc} \cdot \hat{e}_x \psi_{\mu\nu}(x,y,z = 0^+) r\, dr\, d\theta \quad (a)$$

$$b_{\mu\nu} = \frac{1}{2\eta_{\mu\nu}} \int_{x\ section} \mathbf{E}_{inc} \cdot \hat{e}_y \psi_{\mu\nu}(x,y,z = 0^+) r\, dr\, d\theta \quad (b) \quad (5\text{-}113)$$

As an example, let's consider a plane polarized uniform normal plane wave excitation. At the plane $z = 0$, this excitation would have the form $\mathbf{E}_{inc} = E_0 \hat{e}_p$, where $\hat{e}_p$ is a unit normal in the plane wave's polarization direction. An immediate observation is that the excitation just inside the fiber will have the same polarization as the excitation incident. From the modal noise discussion, however, we know that this polarization will not last long. The main point to be made here, however, is that many of the modes of

Equation (5-113) have odd symmetry. As the excitation is even in the case described above, none of these modes will be excited, a fact that is expressed by the relation

$$a_{\mu,2j+1} = b_{\mu,2j+1} = 0 \quad \text{for} \quad j = 0,1,2, \ldots \tag{5-114}$$

with the attendant conclusion that monochromatic excitations are pretty inefficient at coupling power to multimode fibers. Try to find a better excitation than a uniform one.

The immediate question that arises is how efficiently one can couple energy into a single mode fiber. We define a coupling efficiency $F$ for a plane polarized incident disturbance $\mathbf{E}_{inc}$ by the relation

$$F = \frac{\int E_{inc}(x,y,z=0^+) \, \psi_{00}(x,y,z=0^+) \, d^2x}{[\int E_{inc}^2(x,y,z=0^+) \, d^2x]^{1/2} \, [\int \psi_{00}^2(x,y,z=0^+) \, d^2x]^{1/2}} \tag{5-115}$$

The right-hand side of Equation (5-115) is just the left-hand side of the Cauchy-Schwartz inequality and is obviously maximized for $E_{inc}(x,y,z=0^+)$ having the same shape as $\psi_{00}$. This is very nice in that there is a maximum. However, perusal of Equations (5-44) indicates the $\psi_{00}$ is in the form of a Gaussian with a spot size of a few microns. This is to say that the launching system necessary to generate the optimal excitation must have submicron tolerances if it is to come close to achieving maximum efficiency coupling. This is no simple task.

We now wish to consider the problem of incoherent excitation. We again refer to Figure 5.38 for geometrical considerations, and actually the field picture is somewhat correct, at least for single realizations of an ensemble. One thing that clearly does go over is the Fresnel argument that the distribution incident will have nearly the same shape at $z = 0^+$ as at $z = 0^-$. Beyond this, it is best to take a ray approach to solve the problem, as was discussed in the last section. In radiometry, one discusses objects of low coherence in terms of a specific intensity function $I(\mathbf{r}, \Omega)$, which represents the intensity of a ray emanating from a point of coordinate $\mathbf{r}$ on the source (it does not have to be the actual source but a plane in space), pointing in a direction $\Omega$, which is specified by an angle $\theta$ with the normal and an azimuth (see Figure 5.39). We have already seen that inside the fiber the specific intensity must take on a simple form, as for each ray the angle $\theta(z)$ and position $r(z)$ are related through the invariant $R$ parameter (5-99). Therefore, within the fiber the specific intensity must be expressible as a modal distribution function $p(R)$, according to

$$I(\mathbf{r}, \Omega)|_{\text{fiber}} = p(R) \tag{5-116}$$

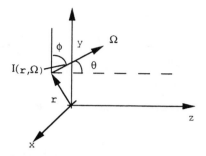

**FIGURE 5.39.**    Parameters important in representing the specific intensity $I(r,\Omega)$ which here is defined for the plane $z = 0$.

A truly incoherent source is not only of infinite extent but is also Lambertian in nature and therefore radiates into a $\cos \theta$ pattern, expressed by

$$I(r,\Omega)|_{\text{incoherent}} = I_0 \cos \theta \qquad (5\text{-}117)$$

As only rays at small $\theta$ are trapped in a fiber, from Equation (5-116) an incoherent source will equally excite all modes of a multimode fiber. The efficiency is still not too good, as the source is infinite, but practical sources like LED's are of finite extent ($\sim 100\ \mu$m) and radiate like $\cos^n \theta$ where $n \geq 1$, as is expressed by

$$I(r,\Omega)|_{\text{LED}} = I_0 \cos^n \theta \ \text{circ} \left(\frac{r}{a}\right) \qquad (5\text{-}118)$$

where the circ function is defined by

$$\text{circ}(r) = \begin{cases} 1, & r \leq 1 \\ 0, & r > 1 \end{cases} \qquad (5\text{-}119)$$

Such a source can therefore be selected to excite a multimode fiber quite efficiently.

The question now arises as to how efficiently an incoherent source can excite a single mode fiber. The diffraction limit, which gives the minimum spot size $a_m$ achievable for a monochromatic source focused down through an optical system of numerical aperture NA is

$$a_m \geq \frac{\lambda}{2\ \text{NA}} \qquad (5\text{-}120)$$

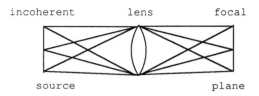

incoherent          lens          focal

source                            plane

**FIGURE 5.40.**    A sketch of an optical system trying to focus down an incoherent source.

The spot size of a single mode fiber is roughly equal to the fiber radius, and for a fiber to be single moded its radius must, from the relation $V \leq 2.4$, satisfy

$$a \gtrsim \frac{\lambda}{2\,NA} \qquad (5\text{-}121)$$

The point is that a single mode fiber is a diffraction limited optical system. The problem with coupling from an incoherent source is, therefore, that the diffraction limit was derived for a spatially coherent source, and this was a necessary condition. As illustrated in Figure 5.40, one cannot focus down a finite incoherent source due to the extent of its numerical aperture. Had the input to the lens of Figure 5.40 been a plane wave (spatially coherent excitation), the system would have been diffraction limited. The only way to get coherence out of an incoherent source is to stick a pinhole of diameter equal to one coherence length over it. But then one is back to coherent excitation for, for example, a single mode.

It is hoped that these considerations, from modal noise through coupling, have convinced the reader that there is some truth to the maxim, "single mode sources with single mode fibers, multimode sources with multimode fibers."

## 5.6 DISPERSION

Consider propagation in a single mode waveguide, a problem depicted in Figure 5.41. What we wish to find is an expression for the envelope function $f(z,t)$ at the plane $z = L$ in terms of the packet $f(0,t)$ that was launched into the fiber. The relations between the scalar fields $E(0,t)$ and $E(z,t)$ and the packet functions $f(0,t)$ and $f(z,t)$ are given by

$$E(0,t) = f(0,t)e^{-i\omega_0 t} \qquad \text{(a)}$$

$$E(z,t) = f(z,t)e^{i\beta(\omega_0)z}e^{-i\omega_0 t} \qquad \text{(b)} \quad (5\text{-}122)$$

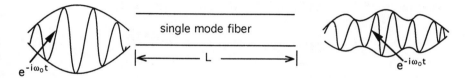

**FIGURE 5.41.** Distortion suffered by an incident wave packet after propagation along a length $L$ of single mode fiber.

where the $\beta(\omega_0)z$ term gives the change in phase between envelope and carrier as the propagation takes place. The spectrum of the wave, which is a constant of propagation due to the linearity and the assumed homogeneously distributed loss of the medium, is defined by

$$E(\omega) = \int E(0,t')e^{i\omega t'} \, dt' \tag{5-123}$$

where the constancy of the spectrum implies that the field at $z$, $E(z,t)$, can be expressed as

$$E(z,t) = \frac{1}{2\pi} \int_{-\infty}^{\infty} E(\omega)e^{i\beta(\omega)z}e^{-i\omega t} \, d\omega \tag{5-124}$$

Note that, from (5-124), if $\beta$ were independent of $\omega$, $E(z,t)$ would simply be $E(0,t)$ multiplied by $e^{i\beta z}$ and there would therefore be no dispersion. This is indeed the case in propagation in a homogeneous medium.

To find our desired relationship, we substitute (5-122b) into (5-124) to obtain

$$f(z,t) = \frac{1}{2\pi} \int_{-\infty}^{\infty} d\omega \, E(\omega)e^{-i(\omega-\omega_0)\left[t-\frac{\beta(\omega)-\beta(\omega_0)}{\omega-\omega_0}z\right]} \tag{5-125}$$

Defining a Fourier transform $\mathscr{F}_{(\omega-\omega_0)}$ by

$$F(\omega - \omega_0) = \mathscr{F}_{(\omega-\omega_0)}(f(t)) = \int_{-\infty}^{\infty} e^{i(\omega-\omega_0)t'}f(t') \, dt' \tag{a}$$

$$f(t) = \mathscr{F}_{(\omega-\omega_0)}^{-1}(F(\omega - \omega_0))$$

$$= \frac{1}{2\pi} \int_{-\infty}^{\infty} e^{-i(\omega-\omega_0)t'}F(\omega - \omega_0) \, d(\omega - \omega_0) \tag{b} \quad (5\text{-}126)$$

one can use (5-122a) and (5-123) in (5-125) to show that

$$f(z,t) = f(0,t) * \mathscr{F}^{-1}_{(\omega-\omega_0)} e^{i\left[\frac{\beta(\omega)-\beta(\omega_0)}{\omega-\omega_0}\right]z} \qquad (5\text{-}127)$$

where $\tau$ is defined by

$$\tau = \frac{\partial\beta}{\partial\omega} \qquad (5\text{-}128)$$

and the * denotes convolution, defined by

$$f(t) * g(t) = \int_{-\infty}^{\infty} f(t-t')g(t')\,dt' = \int_{-\infty}^{\infty} g(t-t')f(t')\,dt' \qquad (5\text{-}129)$$

Consider a simple example that is sufficiently interesting to describe multimode dispersion. Say that $\beta$ is of the form

$$\beta(\omega) = c_1 + (\omega - \omega_0)\,\tau \qquad (5\text{-}130)$$

where $c_1$ is a constant. Using (5-130) in (5-125), one obtains

$$f(z,t) = f(0, t - \tau z) \qquad (5\text{-}131)$$

which states that the packet is delayed as it propagates down the fiber, but its shape is not distorted. In a multimode fiber, however, the input wave is

$$E(0,t) = \sum_{\mu\nu} f_{\mu\nu}(0,t)e^{-i\omega_0 t} \qquad (5\text{-}132)$$

and therefore the output wave packet will be of the form

$$f(z,t) = \sum_{\mu\nu} f_{\mu\nu}(0, t - \tau_{\mu\nu}z) \qquad (5\text{-}133)$$

and the distortion in a three-mode fiber may appear as depicted in Fig. 5.42.

It can be of interest here to see how the pulse distorts as a function of profile. To do this, we need first to find $\beta_{\mu\nu}$ for an $\alpha$ profile as was defined in Equation (5-24). This is really not such a hard task given what has already been accomplished. By making the transformation $x = \ln(r/a)$ one can transform Equation (5-31) to

$$\frac{\partial^2\psi(x)}{\partial x^2} + G^2(x)\psi(x) = 0 \qquad (5\text{-}134)$$

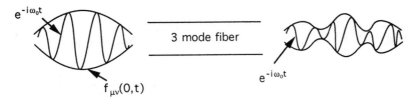

**FIGURE 5.42.**    Possible evolution of a wave packet in a multimode fiber.

Assuming that $G(x)$ is slowly varying, that is, that

$$\left| \frac{\partial G}{\partial x} \psi \right| \ll |G^2(x)\psi(x)| \tag{5-135}$$

one can show that (5-134) has solutions

$$\psi(x) = Ae^{\pm i\int G(x') \, dx'} \tag{5-136}$$

which, by transforming back to the $r$ coordinate, is expressible as

$$\psi(r) = Ae^{\pm i\int k_r(r') \, dr} \tag{5-137}$$

where $k(r')$ is given by Equation (5-103). As the $k_r(r)$ is defined by a square root, its value may be either real or imaginary. The situation is as depicted in Figure 5.43. One can encounter situations in which the $\psi$ has oscillatory behavior throughout the center portion of the fiber and has exponential behavior outside of a turning point, which is defined as an $r$ value for which $k_r(r)$ has a zero. There also can be "helical" modes, for which the field value is zero at the center and that therefore have two turning points. A graphical technique, as is illustrated in Figure 5.44, can be used to determine the turning points and therefore the mode structure. What we are really inter-

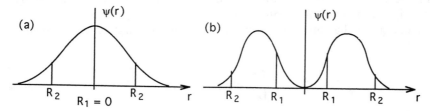

**FIGURE 5.43.**    Forms of the wave functions and the associated turning points for modes which correspond to (a) axial rays and (b) helical rays.

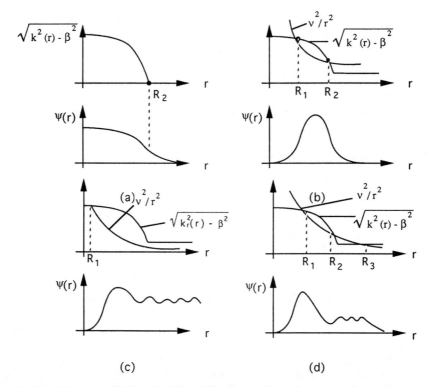

**FIGURE 5.44.** Graphical method for solving for turning points, where the mode structure is plotted below the graphical solution for (a) an "axial" mode, (b) a "helical" mode, (c) a cladding mode, and (d) a leaky mode.

ested in here, though, is not how to determine which modes are leaky but to find the propagation constants.

As we recall from the slab waveguide problem and, more specifically, Figure 2.14, a mode solution will have a $\gamma l$ value of close to an integral multiple of $\pi$. The corresponding quantity here is $\int k_r(r')\,dr'$, and therefore we can guess that the mode condition can approximately be stated in the form

$$\int_{R_1}^{R_2} k(r')\,dr' = \mu\pi \tag{5-138}$$

where $R_1$ and $R_2$ are the turning points. With reference to Figure 5.21, one notices that the total number of modes $M(\beta_c)$ with $\beta$ values greater than a cutoff value $\beta_c$ can be expressed in the form

$$M(\beta_c) = 4 \sum_{v=0}^{v(\beta_c)} \mu(v) \tag{5-139}$$

where the 4 comes from mode degeneracy and $\mu(v)$ is $\mu$ as a function of $v$. Combining (5-139) with (5-138) and passing to the mode continuum limit, one finds that

$$M(\beta_c) = \frac{4}{\pi} \int_0^{v(\beta_c)} \int_{R_1}^{R_2} k_r(r) \, dr \, dv \tag{5-140}$$

Changing the orders of integration and using the notation that $k_0 n(R) = \beta_c$, one finds that

$$M(\beta_c) = \frac{4}{\pi} \int_0^a \int k_r(r) \, dv \, dr = \frac{\alpha}{\alpha+2} \frac{V^2}{2} \left[ \frac{n_1^2 k_0^2 - \beta_c^2}{2 \, \Delta k_0^2 n_1^2} \right]^{\frac{\alpha+2}{\alpha}} \tag{5-141}$$

When $\beta^2 = (1 - 2\Delta)n_1^2 k_0^2$, $M(\beta)$ must be $N$, the total number of modes in the fiber. Using this and solving (5-141) for $\beta$, one finds the relation

$$\beta = n_1 k_0 \left[ 1 - 2\Delta \left( \frac{M^2}{N} \right)^{\alpha/\alpha+2} \right]^{1/2} \tag{5-142}$$

where $M$ should be equivalent to the square of the principal mode number, given by

$$M = 2\mu + v + 1 \tag{5-143}$$

To discuss dispersion, one needs to find the $\tau = \dfrac{d\beta}{d\omega}$ value. In the continuum limit, as in Equation (5-142), this is easily done and the result is

$$\tau = \frac{LN_1}{c} \left[ 1 + \Delta \frac{\alpha - 2 - \epsilon}{\alpha + 2} \left( \frac{m^2}{N} \right)^{\frac{\alpha}{\alpha+2}} + \frac{\Delta^2}{2} \frac{3\alpha - 2 - 2\epsilon}{\alpha + 2} \left( \frac{m^2}{N} \right)^{\frac{2\alpha}{\alpha+2}} + \cdots \right] \tag{5-144}$$

where an expansion in $\Delta$ has been made, and the $N$ and $\epsilon$ are defined by

$$N_1 = n_1 - \lambda \frac{dn_1}{d\lambda} \tag{a}$$

$$\epsilon = -\frac{2n_1\lambda}{n_1\Delta}\frac{d\Delta}{d\lambda} \qquad \text{(b)} \quad \text{(5-145)}$$

Perhaps the best way to comprehend the implications of (5-144) is to consider a pair of examples of different profiles. For example, when $\alpha \rightarrow \infty$ and therefore the profile becomes a step index, $\tau$ can be shown to be given by

$$\tau \cong \frac{N_1}{c}\left(1 + \Delta\left(\frac{m^2}{N}\right)^{1/2} + \frac{\Delta^2}{2}\left(\frac{m^2}{N}\right)^2\right) \qquad \text{(5-146)}$$

Perhaps the most important things to note about (5-146) are that the dispersion is first order in $\Delta$ but that, for lower order modes where $M/N \ll 1$, the dispersion reduces to $N_1/c$, the relationship for a homogeneous medium of index $n_1$. This is clearly due to the fact that the lower order modes do not see the cladding. For $\alpha = 2$, the $\tau$ becomes

$$\tau \sim \frac{N_1}{c}\left[1 + \Delta\frac{\epsilon}{4}\left(\frac{m^2}{N}\right)^{1/2}\right] \qquad \text{(5-147)}$$

which, for small $\epsilon$, is much smaller than the $\tau$ value given in Equation (5-146). In fact, if $\epsilon \rightarrow 0$ — i.e. if different dopants had the same dispersion characteristics — then the dispersion would be second order. One could argue that, if the profile could have an $\alpha$ of $2 + \epsilon$, we could still have second order dispersion. The argument is flawed, however, as $\epsilon(\lambda)$ is a function of $\lambda$. To set $\alpha = 2 + \epsilon$ would require monochromatic illumination, which we have already seen is a bad idea in a multimode fiber. The practical result is that group velocity differential for a parabolic index fiber has a value of roughly 1 GHz km, which says that one propagates a 1 GHz modulation signal for 1 km, a 500 MHz signal for 2 km, etc.

The bandwidth limitation on multimode fibers is the reason that single mode fiber is the transmission medium of choice for the telecommunications industry. As will soon be seen, single mode fiber can have $\tau$'s of 1 psec/nm-km, which implies that for a 1 Å linewidth source, 10 THz could be propagated for a kilometer. In the telephone network (as will be discussed in more detail in the beginning of the next chapter), the long-haul connections require as high a length bandwidth product as possible. This requirement leads to a tremendous single mode fiber market. This is not to say that multimode fiber applications do not remain, however, in applications where coupling efficiency and cost-effectiveness are the major requirements. These applications include data communications and sensor readout. But now our attention will turn to single mode dispersion.

The main equation of single mode dispersion is Equation (5-127), where $\beta(\omega)$ must minimally be expanded to

$$\beta(\omega) = \beta(\omega_0) + (\omega - \omega_0)\frac{d\beta}{d\omega} + \left(\frac{\omega - \omega_0}{2}\right)^2 \frac{d^2\beta}{d\omega^2} + \cdots \quad (5\text{-}148)$$

where, momentarily, we shall see that in a given operating region we actually need the next term beyond that expanded in (5-148). This is the reason why single mode dispersion is so low. One might recall that the index of refraction of a material as a function of frequency might appear as in Figure 5.45 and therefore can be expressed in the form

$$(n^2 - 1) = \sum_{j=1}^{P} \frac{e^2 N_j}{\epsilon_0 m_j(\omega_j^2 - \omega^2)} \quad (5\text{-}149)$$

where $N_j$ is a density and $m_j$ is a mass. Equation (5-149) is often referred to as a Sellmeier equation. In fused silica, the important resonances lie at $\lambda_1 = 0.1\ \mu m$ and $\lambda_2 = 9\ \mu m$, leading to the approximate Sellmeier relation

$$n^2 - 1 = \sum_{j=1}^{2} \frac{\lambda^2 B_j}{\lambda^2 - \lambda_j^2} \quad (5\text{-}150)$$

where the $B_j$ are empirically determined. The $n(\lambda)$ and $d\tau/d\lambda$ for bulk fused silica are sketched in Figure 5.46. The main point here is that the $d\tau/d\omega = d^2\beta/d\omega^2$ has a zero at 1.27 $\mu m$, and therefore the first term is the dispersion equation would be third order in $(\omega - \omega_0)$. This is the reason why dispersions of 1 psec/nm-km are achievable.

Figure 5.46 is not quite the whole story. Going to single mode fiber eliminated modal dispersion but not the material dispersion. Operating at the minimum should eliminate the material dispersion but not the wave-

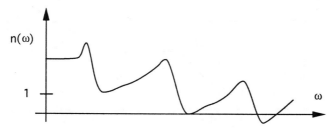

**FIGURE 5.45.**    Possible index of refraction as a function of frequency.

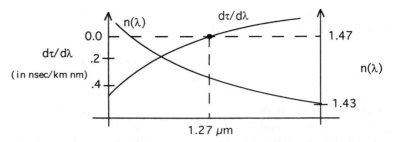

**FIGURE 5.46.** Variations of $n(\lambda)$ and $d\tau/d\lambda$ for bulk fused silica about the dispersion minimum at $\lambda = 1.27\ \mu m$.

guide dispersion, which comes about due to the fact that $d\tau/d\lambda$ will be different in the core and the cladding. To conceptualize this, one could use $2/V$ for $M^2/N$ in Equation (5-144) to obtain

$$\tau \approx \frac{N_1}{c}\left[1 + \frac{\Delta}{3}\left(\frac{a_c}{a}\right)^2 + \cdots\right] \tag{5-151}$$

where $a_c$ is the cutoff radius, to note that $\tau$ will be quite sensitive to the fiber radius $a$. Plots of $d\tau/d\lambda$ for various values of $a$ are illustrated in Figure 5.47. Changing the radius has the effect of shifting the dispersion minimum to longer wavelengths. This is a favorable characteristic, as the loss minimum (see Figure 5.25) occurs at about 1.55 $\mu m$. As it turns out, though, small values of $a$ simply complicate the fabrication and coupling problems. A more practical solution to dispersion shifted fiber has been the use of the so-called W fiber. As is illustrated in Figure 5.48, the basic idea behind a W fiber is to include more parameters in the fiber so as to be able to set the minimum wherever necessary.

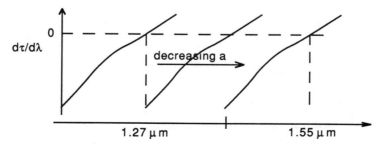

**FIGURE 5.47.** Plots of $d\tau/d\lambda$ for various values of the radius $a$.

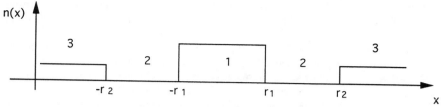

**FIGURE 5.48.** Profile of a W fiber.

# References

Alferov, Zh. I., M.V. Andreev, E.L. Portnoi and M.K. Trukan, 1970. "AlAs-GaAs Heterojunction Injection Lasers with a Low Room-Temperature Threhold." *Sov. Phys. Semiconductors* **3,** 1107–1110.

Barrow, W.L., 1936. "Transmission of Electromagnetic Waves in Hollow Tubes of Metal." *Proc. IRE* **24,** 1298–1328.

Basov, N.G., O.N. Krokhin and Yu.M. Popov, 1961. "Production of Negative Temperature States in P-N Junctions of Degenerate Semiconductors," *J.E.T.P.* **40,** 1320.

Carson, J.R., S.P. Mead and S.A. Schelkunoff, 1936. "Hyper-Frequency Wave Guides—Mathematical Theory." *Bell Sys. Tech. J.* **15,** 310–333.

Eriksrud, M. and A.R. Mickelson, 1982a. "Experimental Investigation of the Variation of Backscattered Power Level with Numerical Aperture in Multimode Optical Fibers." *Elect. Lett.,* **18,** 130–132.

Eriksrud, M. and A.R. Mickelson, 1982b. "Application of the Backscattering Technique to the Determination of Parameter Fluctuations in Multimode Optical Fibers." *IEEE J. Quant. Elect.,* **JQE-18,** 1478–1483; *IEEE Trans. Microwave Theory Tech.,* **MTT-30,** 1466–1471.

Hall, R.N., G.E. Fenner, J.O. Kingsley, T.J. Soltys and R.O. Carlson, 1962. "Coherent Light Emission from GaAs Junctions." *Phys. Dev. Lett.,* **9,** 366.

Hayashi, I and M.B. Panish, 1970. "GaAs-Ga$_x$ Al$_{1-x}$As Heterostructure Injection Lasers Which Exhibit Low Thresholds at Room Temperature." *J. Appl. Phys.,* **41,** 150–163.

Hjelme, D.R. and Mickelson A.R., 1983. "Microbending and Modal Noise." *Appl. Opt.* **22,** 3874–3879.

Hondros, D. and P. Debye, 1910. "Elektromagnetische Wellen an Dielektrischen Drähten." *Ann. Phys.,* **32,** 465–476.

Jones, M.W. and K.C. Kao (1969) "Spectrophotometric Studies of Ultra Low Loss Glasses." *J. Phys. E.,* **2,** 331–335.

Kapron F.P., D.B. Keck and R.D. Maurer, 1970. "Radiation Losses in Glass Optical Waveguides." *Appl. Phys. Lett.,* **17,** 423–425.

Kurtz, C.N. and W. Streifer, 1969. "Guided Waves in Inhomogeneous Focusing Media Part I: Formulation Solution for Quadratic Inhomogeneity." *IEEE Trans. Microwave Theory Tech.,* **MTT-17,** 11–15.

Maimon, T.H., 1960. "Optical and Microwave-Optical Experiments in Ruby." *Phys. Rev. Lett.,* **4,** 564–565.

Mickelson, A.R. and M. Eriksrud, 1981. "Role of Modal Distribution in Determining Power Backscattered from Fibers with Diameter Perturbations." *Elec. Lett.,* **17,** 658–659.

Mickelson, A.R. and M. Eriksrud, 1982. "The Theory of the Backscattering Process in Multimode Optical Fibers." *Appl. Opt.,* **21,** 1898–1909.

Nathan, M.I., W.P. Dumke, G. Burns, F.H. Dills and G. Lasher, 1962. "Stimulated Emission of Radiation from GaAs *p-n* junctions." *Appl. Phys. Lett.,* **1,** 62.

Schriever, O., 1920. "Elektromagnetische Wellen an Dielektrischen Drähten." *Ann. Phys.,* **64,** 645–673.

Snitzer, E., 1961. "Cylindrical Dielectric Waveguide Modes." *J. Opt. Soc. Amer.,* **51,** 491–498.

Snyder, A.W. and W.R. Young (1978). "Modes of Optical Waveguides." *J. Opt. Soc. Am.,* **68,** 297–309.

Southworth, G.C., 1936. "Hyper-Frequency Wave Guides—General Considerations and Experimental Results." *Bell Sys. Tech. J.,* **15,** 284–309.

Yadlowsky, M.J., 1992. "Time-Dependent Radiative Transfer in Waveguides and Its Application to the Measurement and Characterization of Multimode Fiber Systems." Ph.D. Dissertation, University of Colorado, Boulder.

Zahn, H., 1916. "Über den Nachweis Elektromagnetichen Wellen an Dielektrischen Drähten." *Ann. Phys.,* **49,** 907–933 (1916).

## PROBLEMS

1. Consider an index distribution, where $n_1 > n_2 > n_3$, as depicted in Figure 5.49. Try to solve for the exact modes of this distribution by treating the propagation problem as a boundary value problem.
   (a) What solutions should be assumed in each of the regions I, II, and III?
   (b) Write down the continuity equations across the boundaries.
   (c) What are the limits on the values of $\beta$ allowable for guided modes?
   (d) Sketch the intensity patterns of the first few modes.
2. Solve the following:
   (a) By going through the transformations indicated in Equations (5-34)–(5-37) in the chapter, derive in some detail Equation (5-37) from Equation (5-33).
   (b) Find the recurrence relation from the $L$'s of Equation (5-37) and

**FIGURE 5.49.**

thereby show that they must all be integers if they are to be zero after a finite number of terms.

(c) Find the set of equations that would hold at $r = a$ to determine $\beta$, if one were not to assume that the parabolic index profile were infinite. Recall that four equations will be necessary to completely determine the problem.

(d) In the case of (c) above, what will the modal polarizations look like? Will they still be LP? Will they be hybrid?

3. In the following, you are to plot the evolution of the tip of the polarization vector for several different values of the phase difference $\Delta\beta z = (\beta_1 - \beta_2) z$ for several different values of the angle $\theta$, for the following mode combinations:

(a) $\psi_1 = \hat{e}_x, \quad \psi_2 = \hat{e}_y$
(b) $\psi_1 = \hat{e}_x + i\hat{e}_y, \quad \psi_2 = \hat{e}_x - i\hat{e}_y$
(c) $\psi_1 = \hat{e}_x e^{i\phi}, \quad \psi_2 = \hat{e}_y e^{-i\phi}$
(d) $\psi_1 = \hat{e}_x \cos\theta + \hat{e}_y \sin\theta, \quad \psi_2 = -\psi_2 = -\hat{e}_x \sin\theta + \hat{e}_y \cos\theta$
where the state vector $\psi = \text{Re}[(\psi_1 e^{i\beta_1 z} + \psi_2 e^{i\beta_2 z})e^{-i\omega t}]$

4. Consider the mode patterns depicted in Figure 5.50. Consider that the $\beta$'s of these four modes become split due to some perturbation induced in the fiber during the fabrication process. Sketch the evolution of the modal polarization as a mode propagates from the input, given the following excitations:

(a) The $HE_{21}$ even and $TM_1$ excited in phase
(b) The $HE_{21}$ odd and $TE_1$ excited in phase

Can you recognize any of the resulting modal polarizations in terms of LP modes?

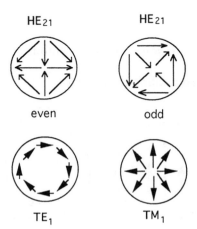

FIGURE 5.50.

5. As discussed in the text, the LP modes are really made up of linear combinations of hybrid modes, at least in a perfect fiber. The propagation constants of these hybrid modes, were one to calculate to higher order in the index contrast, would in general be distinct. Therefore, an originally excited LP would exhibit polarization evolution in propagating down the fiber. Make numerous sketches of the evolution of the polarization states of the $\mathbf{LP_{00}}$, $\mathbf{LP_{01}}$, $\mathbf{LP_{02}}$, and $\mathbf{LP_{10}}$ modes as they propagate down a "perfect," "strongly" guiding fiber.

6. Consider the fiber modes depicted in Figure 5.51 as the only two that can propagate in a given, almost single-mode fiber. Say we imagine in a diffraction limited manner (that is, the imaging system has a much higher NA than the fiber) through a polarizer onto a detector of the size of the diffraction limited spot, which is given by $\dfrac{\lambda}{2NA}$, where NA is the NA of the fiber, as depicted in Figure 5.52.
   (a) What is the power output of the detector if only mode 1 is propagating in the fiber?
   (b) What is the power output of the detector if only mode 2 is propagating in the fiber?
   (c) Repeat (a) and (b) with the polarizer removed.

7. Say that the evolution of polarization in an elliptical fiber is given by

$$\frac{da_x}{dz} = i\kappa e^{i\Delta\beta z} a_y$$

$$\frac{da_y}{dz} = i\kappa e^{-i\Delta\beta z} a_x$$

where $\Delta\beta = \beta_x - \beta_y$.
   (a) Solve these equations for the evolution of $a_x$ and $a_y$. Assume $\Delta\beta \sim 2\kappa$.
   (b) What does the solution look like (draw a picture) when $\Delta\beta \to 0$?

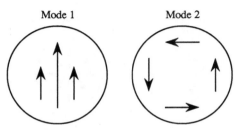

**FIGURE 5.51.**

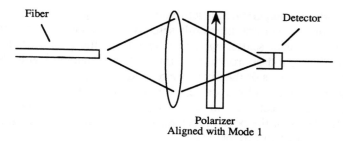

Fiber

Detector

Polarizer
Aligned with Mode 1

**FIGURE 5.52.**

(c) What happens to the amount of coupling between the states as $\Delta\beta$ increases? When is the coupling a minimum? Why?

(d) What problem which you had earlier does this remind you of? Which quantities were analogous there?

8. Solve the following:

(a) Rederive Equation (5-59) of the chapter in some detail, but this time assuming that $\epsilon_{yx} = \epsilon_{xy}^*$ in the first part of Equation (5-51). Do the resulting equations conserve energy?

(b) Find a general solution, comparable to Equation (5-61) of the chapter, to the analog to Equation (5-59) of the chapter found in (a).

(c) Transform the equations found in (a) to a circular basis, that is, a set of equations for $a_r = a_x + ia_y$ and $a_l = a_x - ia_y$ where $\Delta\beta = \beta_r - \beta_l$.

(d) Find a general solution to the equations of (c). How does this solution compare to that found in (b)?

9. Recall that the coupled polarization equations look like

$$\frac{da_x}{dz} = i\chi \, a_y \, e^{i\Delta\beta z}$$

$$\frac{da_y}{dz} = i\chi \, a_x \, e^{-i\Delta\beta z}$$

(a) Solve these equations exactly for $a_x(z)$ and $a_y(z)$ in terms of $a_x(0)$, $a_y(0)$ for $\chi$ and $\Delta\beta$ constants in $z$. Show your work.

(b) Take the limit where $\Delta\beta \to 0$. How might one "fix-up" this solution to take into account that $\chi$ is a function of $z$? Can you estimate the "decoupling length" for the polarization in terms of the standard deviation of the zero mean Gaussian process $\chi$?

(c) Take the limit $\chi \to 0$. What happens to this result if $\Delta\beta$ is $z$ varying? Why?

**10.** Consider the experimental setup depicted in Figure 5.53. Say that the detector/receiver are sufficiently sensitive to register the time evolution of the Rayleigh scattering in the fiber from the (assumed delta-like) periodic (period $\gg$ round trip time in fiber) incident pulses. Say that scattering factors of the fibers are related by $\alpha_{R3} < \alpha_{R1} < \alpha_{R2}$, and the absorption factors by $\alpha_{a2} < \alpha_{a1} < \alpha_{a3}$. The splices have $\sim 0.1$ dB loss, with some reflection, and the termination point is polished. Make a detailed plot of the power received as a function of time, assuming the fibers are all 1 km in length.

**11.** Say we have the two modes

$$\psi_1 = \sqrt{\frac{2\eta_1}{a}} \cos\frac{\pi}{2a} x e^{i\left(\sqrt{k^2 - \frac{\pi^2}{4a^2}}\right)z}$$

$$\psi_2 = \sqrt{\frac{2\eta_2}{a}} \sin\frac{\pi}{a} x e^{i\left(\sqrt{k^2 - \frac{\pi}{a^2}}\right)z}$$

propagating in a strongly guiding slab waveguide of width $2a$, extending from $-a < x < a$. Give expressions for and sketch the interference patterns of the modes if they are equally excited for the following $z$ values:

**(a)** $z = 0$

**(b)** $z = \dfrac{\pi}{2\Delta\beta}$

**(c)** $z = \dfrac{\pi}{\Delta\beta}$

**(d)** $z = \dfrac{3\pi}{2\Delta\beta}$

where $\Delta\beta = \sqrt{k^2 - \dfrac{\pi^2}{4a^2}} - \sqrt{k^2 - \dfrac{\pi^2}{a^2}}$

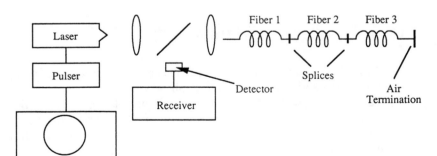

**FIGURE 5.53.**

12. We wish to consider modal noise in a slab waveguide with an index of
   $n_1 = 1$ for $-a < x < a$, and $n_2 \rightarrow -\infty$ for $|x| > a$. Do not use the weakly
   guiding approximation.
   (a) Find the normalized propagating modes of the structure. How
       many modes are propagating for $a = 2\lambda$?
   (b) Say the phases of the propagating modes are randomized by ther-
       mal variations. Find the mean and standard deviation of the field
       value at a point $b$ in the core for $a = 2\lambda$ and for $a \rightarrow \infty$.
   (c) Find the mean and standard deviation of the power received by a
       detector of circular area $\pi b^2$ centered on the fiber axis, for $b$ varying
       from 0 to $a$ and for $a = 2\lambda$ and $a \rightarrow \infty$.
13. The statistics of modal noise in a fiber can be described by the Beta
   distribution

$$f_f(P|P_T) = \frac{\Gamma(M_T)}{\Gamma(M)\Gamma(M_T - M)} \frac{1}{P_T} \left(\frac{P}{P_T}\right)^{M-1} \left(1 - \frac{P}{P_T}\right)^{M_T - M - 1}$$

   where $f_f(P|P_T)$ gives distribution of measured power values (i.e.,
   $\bar{P} = \int dP f_f(P|P_T)$) given that one receives $M$ modes, carrying power $P$ of
   a total of $M_T$ modes representing a power $P_T$.
   (a) Show that $f_f$ reduces to the gamma distribution $f_f(P)$

$$f_f(P) = \frac{1}{\Gamma(M)} \left(\frac{M}{\bar{P}}\right)^M P^{M-1} e^{-MP/\bar{P}}$$

   in the limit that $M \rightarrow 1$. Plot this function.
   (b) Calculate and plot the beta distribution for $M = M_T/2$, assuming
       large $M_T$.
   (c) What happens to the beta distribution when $P \rightarrow P_T$?
   (d) What happens when $P_T \rightarrow \infty$?
   Note: One might refer to the treatment in Hjelme and Mickelson
   (1983).
14. Solve the following:
   (a) With reference to Figure 5.38 of the chapter, write down the exact
       equations satisfied by $E_{inc}$, $E_{rad}$ and $E_b$ at $z = 0$ for an inhomoge-
       neous slab with everything polarized in the $y$ (out of the paper)
       direction. Assume the radiation modes can be expressed as

$$E_{rad} = \int_{-\theta_c}^{\theta_c} \psi_r(\theta) e^{ikx\sin\theta} e^{ikz\cos\theta} \, d\theta$$

   (b) Try to obtain a relation involving an overlap integral plus something else. What is the something else? When might it be small? When might it be large?

15. Consider a (weakly guiding) index distribution

$$n^2(x) = n_a^2 \left[ 1 - 2\Delta \left( \frac{x}{a} \right)^2 \right]$$

and recall that the ray equation takes the form

$$\frac{\partial}{\partial s} \left[ n(x, y, z) \frac{\partial \mathbf{r}}{\partial s} \right] = \Delta n$$

and the curl equation takes the form

$$\nabla \times \mathbf{E} = -\frac{\partial \mathbf{B}}{\partial t}$$

$$\nabla \times \mathbf{H} = \frac{\partial \mathbf{D}}{\partial t}$$

(a) Find the paraxial ray equations for this medium.
(b) Solve for the ray paths.
(c) Find the wave equation for this medium and assume a WKB solution, that is, a solution of the form

$$e^{\int \sqrt{k^2(x) - \beta^2} \, dx}$$

to obtain solutions in the guided and nonguided regions.
(d) Use the WKB solutions and BC's to find a relation for determination of the propagation constant.

16. In the chapter, the solutions to the ray equation

$$\frac{\partial}{\partial s} n(\mathbf{r}) \left( \frac{d\mathbf{r}}{ds} \right) = \Delta n$$

in cylindrical coordinates for $\alpha$ index profile fibers were discussed. Find explicit exact expressions for the ray trajectories in a parabolic index fiber, and explicitly evaluate the group delay

$$\tau = \int n \, ds$$

Are there any other profiles you can solve these problems for? If there are, do one.

17. In this problem, we wish to calculate and plot exact ray paths for some strange and interesting profiles, using Equations (5-96), (5-97) and (5-106) of the chapter. The index distributions are as follows:

(a) $n^2(r) = \dfrac{n_1^2}{r^2}$

(b) $n^2(r) = n_1^2$

(c) $n^2(r) = n_1^2 - n_2^2 r^2$

(d) $n^2(r) = n_1^2 - n_2^2 \tan^2 cr$

18. Consider a two-dimensional $(x - z)$ medium which is uniform in $z$ but whose transverse index profile is expressible as

$$n(x) = n_1\left[1 - \frac{\Delta}{s}\left[\left(\frac{x-b}{s}\right)^2 + \left(\frac{x+b}{a}\right)^2\right]\right]$$

In part (a)–(d), one should take $b \ll \dfrac{1}{\sqrt{2\Delta}}$, and $\Delta \ll 1$.

(a) Plot the transverse profile function for various values of $\dfrac{b}{a}$.

(b) Solve the paraxial ray equation for this profile.

(c) Define an $R$ parameter as was done in (5-99).

(d) Plot ray paths for various values of $R$.

(e) Explain what is strange about the answer you found in (b). Can you find a first-order correction for $b \leq \dfrac{a}{\sqrt{2\Delta}}$?

19. In the mode continuum limit, one can write that the power $P$ in a fiber is given by

$$P = \int p(R)m(R)\, dR$$

where $p(R)$ is the modal distribution and $m(R)$ the mode density given by

$$m(R) = V^2 R^{1+\frac{4}{\alpha}}$$

The near field intensity at the end face of the fiber is given by

$$I\left(\frac{r}{a}\right) = \frac{V^2}{\pi}\int_{\left(\frac{r}{a}\right)^{\alpha/2}}^{1} p(R)R\, dR$$

where the profile function of the fiber is given by the $\alpha$ expression

$$n^2(r) = n_1^2 \left[ 1 - 2\Delta \left( \frac{1}{\alpha} \right)^\alpha \right]$$

Find and interpret the power and near-field intensity for the following modal distributions:

(a) $p(R) = A\delta(R - R_0)$
(b) $p(R) = A(1 - qR), \qquad qR < 1$
(c) $p(R) = A \, \text{rect} \left( \dfrac{R}{R_m} \right)$
(d) How might the distributions of (b) and (c) be used to find the fiber profile?

20. Describe in detail how one would use the WKB method in concert with the equivalent index method to find the propagation constant and fields of the lowest order mode of a two-dimensional profile described by $f(x,y)$. Assume the profile has maximum value $f(x=0, y=0)$ and decreases monotonically in both the $x$ and $y$ directions. Check your results with those of Marcatili in the limit that the guide becomes a rectangle.

21. Consider an index profile of the form

$$n^2(r) = \begin{cases} n_1^2 + \Delta n \cos^2 \dfrac{\pi}{\alpha} r & r < \dfrac{3}{2}\alpha \\[2mm] n_1^2 & r < \dfrac{3}{2}\alpha \end{cases}$$

(a) Use Equation (5-137) of the chapter to find the (approximate) forms of solution in each part of the profile.
(b) Sketch the first several modes of this profile, in the form of Figure 5.44.
(c) Sketch out the profile of a leaky mode and a cladding mode.
(d) Will the dispersion of this profile be worse or better than that for a step index one? Explain.

22. Consider an azimuthally symmetric incoherent light source radiating into an azimuthally symmetric optical fiber with index profile $n(r)$. Say that the source can be characterized by a specific intensity $I(r, \sin \theta)$ and the fiber by its acceptance angle $\sin \theta_c(r) = \sqrt{n^2(r) - n_c^2}$.
(a) Find general expressions for the amount of coupled power and the coupling efficiency when the cores are perfectly aligned.

(b) Evaluate these expressions for a uniform source of NA $= \sin \theta_s$ and radius $a_s$ radiating into a step index fiber with NA $= \sin \theta_f$ and radius $a_f$.

(c) Repeat (a) but for the case where the cores are offset by a distance $\delta$.

(d) Try to repeat (b) for the condition of (c).

23. Say we excite a three-mode waveguide with a pulse $e^{-t^2/T^2}$ such that the modes are equally excited. Say the group velocities of the modes are given by

$$V_1 = 2 \times 10^8 \text{ m/sec}$$

$$V_2 = 2.01 \times 10^8 \text{ m/sec}$$

$$V_3 = 2.02 \times 10^8 \text{ m/sec}$$

Sketch the pulse shape after 100 m, 1 km and 10 km for:

(a) $T = 1$ sec

(b) $T = 1$ msec

(c) $T = 1$ $\mu$sec

(d) $T = 1$ nsec

24. In the chapter, it was shown that the envelope function of a wave could be expressed as a convolution of a distortion function with the delayed input packet. Perform the Fourier transform explicitly on the distortion function for $D = d^2\beta/d\omega^2 = $ constant of $\omega$. Peform the convolution explicitly for Gaussian input packets. Plot the result for some "reasonable" parameters.

25. Solve the following:

(a) Derive Equation (5-144) of the chapter.

(b) Plot $\tau$ as a function of $\alpha$ for several given $M/N$ values.

(c) Consider a fiber in which $\alpha$ is a random function of $z$, the length down the fiber, with mean value $\alpha_0$ and standard derivation $\Delta\alpha$. Sketch the evolution of a pulse if $\alpha_0 = 3$ and $\Delta\alpha = .25$. Repeat for $\alpha_0 = 2$ and $\Delta\alpha = 0.2$. Pay attention to the length dependence of the pulse width.

26. Consider a 1 km piece of fiber with a given spec sheet. On the spec sheet we are given a number $D \left( \text{in } \dfrac{\text{Psec}}{nm \text{ km}} \right)$ for the spectral dispersion, a function $\alpha(\lambda)$ (in dB/km) for the spectral extinction and a wavelength operating range $\lambda_{min} < \lambda_{op} < \lambda_{max}$. Consider a pulse to be launched in the fiber with a given spectrum $I(\lambda)$, but so narrow in time ($< 1$ psec) that one can take it to be a delta function at the fiber input. For a single mode fiber

(a) Find an expression for the temporal shape of the pulse.

(b) Consider $D = 50$, $\alpha(\lambda) = \log \left[ \dfrac{\lambda_{max} - \lambda_{min}}{\lambda_{max} - \lambda} \right]$ and $I(\lambda) = 1$ $\lambda_{min} <$ $\lambda < \lambda_{max}$, plot the time evolution of the output pulse.

(c) Consider parameters as in the above, but for a multimode fiber, with intermodal dispersion of 500 psec/km. Plot the output pulse shape in this case.

# 6

---

# Integrated Optics

The purpose of the present chapter is to introduce some of the concepts and techniques, both practical and theoretical, of the field of integrated optics. The first section of the chapter is devoted to trying to explain what is meant here by integrated optics. Depending upon whose definition one wants to use, the field can be of almost infinite extent. As a much more limiting definition is employed here, the second section of the chapter will concentrate on an exposition of the physics and practice of the electrooptic effect. All of the integrated optical components considered in the third section of this chapter will be planar technology fabricated, electrically controlled, optical processing devices. The basic technology considered will be that of $LiNbO_3$ indiffusion, although all the basic principles are transferable to other electrooptic insulator materials as well as electrooptic semiconductors. The third section of the chapter will introduce concepts and applications through examples of specific devices.

## 6.1 THE WHAT AND WHY OF INTEGRATED OPTICS

Generally, when one refers to integrated optics, one refers to a set of electrooptic components which can perform some signal processing function. The signals operated on are generally optical, but the control signals are electrical. As signal processing functions on optical signals generally require either modulation or routing of signals and it is much simpler to switch a single mode than to switch multiple modes, these devices are in general single mode and, further, compatible for use with single mode fibers, a driving force behind integrated optics. Integrated optical components are generally fabricated in electrooptic substrates by methods of planar circuit technology. Although the original integrated optical components were by no means integrated, as there was generally only one component per fiber coupled substrate, the use of planar circuit technology in integrated optic fabrication allows for integrated optics to possibly follow in the footsteps of silicon technology. After an initial period in the 1950s when only single transistors could be placed on a silicon substrate, silicon afterwards passed through the phases of medium scale integration (MSI) and large scale integration (LSI) and is now in a period of very large scale integration (VLSI). Already in 1988, $LiNbO_3$ integrated optics technology has produced marketable circuits with more than 100 individual components on a substrate, enough to qualify in silicon parlance as being MSI. Although it is difficult and risky to try to tell the future, there seems to be no reason to think that circuit density in $LiNbO_3$ and other competing technologies cannot and will not continue to increase in the future. The rate of this increase, however, will more likely than not be determined by market factors.

To get some understanding of the driving forces behind present-day integrated optic development, it is worthwhile to consider what some of the driving forces behind the earlier pushes in integrated optics were. A schematic of an idealized telephone network is given in Figure 6.1. As much as anything else, the point is that the network is hierarchical. At the lowest level is the local loop, in which small numbers (tens) of telephone conversations (at 96 kbps digital rate) are carried. The local branch exchange (LBX) serves as a multiplexer (MUX) for these conversations in order to launch them on a trunk and as a demultiplexer (DEMUX) for the trunk in order to correctly address individual calls for entrance into the loop. The long-haul lines are at

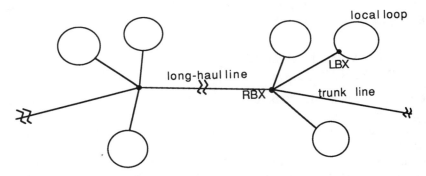

**FIGURE 6.1.**    A schematic depiction of a portion of a telephone network in which long-haul lines join regional branch exchanges (RBX's) which in turn are joined by trunk lines to local branch exchanges (LBX's), which serve as hubs for local (customer) loops.

the top of the hierarchy and, depending on how long the haul, can require rates up into multigigabits. The regional branch exchanges (RBX's) therefore have a truly complicated, high-speed MUX/DEMUX operation to perform.

Not so very long ago (the 1930s and 1940s), the telephone network was essentially analog. Long-haul transmission required multiple coaxial or twisted pair lines. The switch to digital required an order of magnitude leap in rate (digital coding requires roughly twenty times the analog transmission rate) plus an increase in bandwidth in order to pass the pulse edges. The timing of the maturity of the fiber was very fortuitous, as it essentially overlapped the period in which the trunk lines were being switched over from analog to digital. As digital switches became available for LBX's and RBX's, it was as easy to use direct laser modulation digital schemes to drive hundred megabit multimode fibers as it would have been to install multiple coaxes or twisted pairs. As this period in the mid-1970s, these fiber rates were sufficient for trunk capacity. With the advent of the practicality of the high bandwidth single mode fiber, the fiber then entered the long haul. As the RBX's were already digital, it was an easy job to simply string fibers over all sorts of new long-haul routes and at previously undreamed-of rates. However, although the appearance of digital switching and fiber optics at roughly the same time served to promote fiber directly into the network, it left a bit of a strange looking network in that the fiber was being used directly as a replacement for wire. But fibers are not wires, nor are lasers transistor amplifiers.

The places where the fiber/digital electronics hybrid solution appears most forced is in repeaters and RBX's. For example, consider the schematic depiction of a hybrid repeater in Figure 6.2. The repeater needs a detector, a pre-amp, an amplifier, a driver, and a laser to achieve the amplification process. A more recent solution for repeating, the active optical fiber, is finding applicability. However, the erbium gain curve is in the 1.55 $\mu$m window area and most communications today are done at a wavelength of 1.3 $\mu$m. The depiction of the RBX in Figure 6.3 appears as an even more forced solution than the hybrid repeater of Figure 6.2. Much more natural solutions would be those depicted in Figures 6.4 and 6.5. Here the repeating

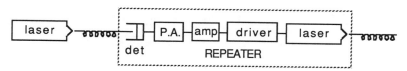

**FIGURE 6.2.**  A portion of a long-haul line containing a repeater which requires complete electronic decode.

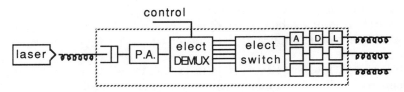

**FIGURE 6.3.** An RBX which requires complete electronic decoding.

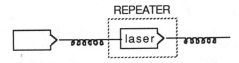

**FIGURE 6.4.** A long-haul line containing an all optical repeater.

is taken care of by using a travelling wave semiconductor laser repeater as a "nonintegrated" but perhaps integrable integrated optic component which is beginning to look more and more practical. A major portion of this chapter will be taken up in considering components of the integrated optical switch which appears in the RBX of Figure 6.5. Although 8 × 8 versions of such switches were already obtainable off the shelf in 1988, the prices of the components are very high.

Although the solutions of Figures 6.4 and 6.5 seem eminently reasonable at present, they did not in the late 1970s and early 1980s when the massive switchover of the telephone network to digital switching and fiber transmission took place. The push in integrated optics (IO), at Bell Labs at least, started in circa 1969 but did not come in time for the fiber/laser/digital switch revolution. It is somewhat unlikely that the solutions of Figures 6.4 and 6.5 are going to find extensive application in the near future if only because a telephone system with over capacity is already in place and would

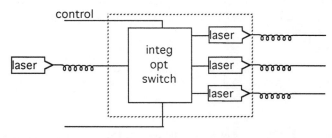

**FIGURE 6.5.** An RBX containing an electrooptic switch and all optical repeaters.

be very expensive to alter at this point. There are other applications, however, that are important to telecommunications as well as to other fields, which could lead to more widespread IO usage. Perhaps the most noteworthy of these is external modulation.

Figure 6.6 illustrates some effects that occur to the laser spectrum under direct modulation. The line formation process causes the spectrum to change dynamically during each modulation period, with a proportionate dispersion penalty during transmission. Equally serious if not more so is the frequency chirp which causes the whole gain packet to move relative to the eventual center wavelength $\lambda_0$. It is easy to see that external modulation of a stabilized linewidth semiconductor laser, contrasted with the direct modulation case in Figure 6.7, will entail a smaller dispersion penalty at a cost of some increased insertion loss. There are a good number of system applications in which this tradeoff is very worthwhile. For other kinds of modulation, only external modulators can be used. Heterodyne detection schemes operate most advantageously with either frequency or phase modulation. Current modulation can never yield pure frequency modulation due to the amplitude frequency coupling, and phase modulation is only achievable with external modulation. Also, polarization modulation can only be achieved externally.

Other applications of IO are appearing every day in the areas of optical signal processing and IO sensing. In optical time domain reflectometry, a

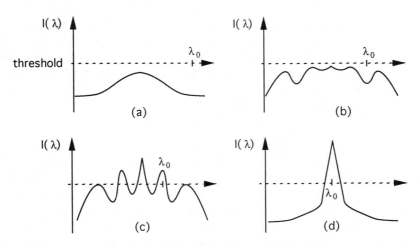

**FIGURE 6.6.**    Dynamics of a laser under transient approach to the steady state from far under threshold: (a) laser operating as an LED, (b) just below threshold where mode structure begins to become evident, (c) just over threshold where a dominant mode begins to appear, and (d) well over threshold and at operating point.

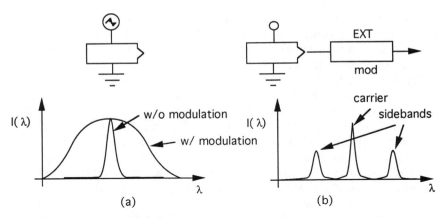

**FIGURE 6.7.**   Lasers under modulation and the spectra generated for (a) direct current modulation and (b) external intensity modulation.

short pulse is launched into a fiber or other optical transmission medium, and the low level backscattered response is monitored as a diagnostic of channel conditions. In cases where the laser is current modulated and the beam is focused into the optical system, any Fresnel reflection from the launching optics tends to swamp the backscattered level and drive the receiving electronics into saturation. A simple chip with two optical switches and some drive electronics to amplify synchronized gating pulses can allow the laser to be run continuous wave (CW) and the pulse to be gated into the receiver after the Fresnel reflection has radiated away. Another application is in the fiber optic gyroscope, where it can be very advantageous to phase modulate one of the counterpropagating beams with respect to the other. The only simple single mode fiber compatible method to do this is an IO solution. Once the IO solution is chosen, another advantage is that additional components can be added to the substrate such that routing and signal mixing can also be carried out.

The case with fiber optics was that the processing and sensing markets did not grow as explosively as the telecommunications market. However, the processing and sensing markets continue to grow at a steady rate, while the telecommunications market growth has slowed significantly. With IO, a telecommunications market did not materialize, whereas the processing and sensing markets are beginning to appear. Phase shifting and modulating devices on the market are beginning to sell, with switching devices selling to a lesser degree. It seems quite likely that these markets should continue to grow in the future.

## 6.2 THE ELECTROOPTIC EFFECT

Figure 6.8 represents a blowup of a portion of a lattice in which there is excess positive charge at lattice sites and in which the electrons are bound at symmetrically located points between the lattice sites. The electrons are bound to their positions by, most probably, covalent bonding forces, which can be represented by a potential energy function $V(\mathbf{r})$. In the simplest model of electron motion, an external force $\mathbf{F}_{ext}$ applied to an electron will result in electron motion described by Newton's law

$$m\ddot{\mathbf{r}} = \mathbf{F}_f + \mathbf{F}_r + \mathbf{F}_{ext} \tag{6-1}$$

where $\mathbf{r}$ is the electron's position, assumed zero at equilibrium, the dots denote differentiation with respect to time, $\mathbf{F}_f$ is a frictional force which comes about due to phonon production, and $\mathbf{F}$ is a restoring force which must be given by

$$\mathbf{F}_r = -\nabla V \tag{6-2}$$

For the very simple isotropic case of Figure 6.8, the $V(\mathbf{r})$ must be given by

$$V(\mathbf{r}) = \frac{1}{2} K_0 r^2 \tag{6-3}$$

and the equation of motion (6-1) reduces to

$$\ddot{x} + \gamma\dot{x} + \omega_0^2 x = F_{ext|x} \tag{a}$$
$$\ddot{y} + \gamma\dot{y} + \omega_0^2 y = F_{ext|y} \tag{b}$$
$$\ddot{z} + \gamma\dot{z} + \omega_0^2 y = F_{ext|z} \tag{c} \tag{6-4}$$

where the $\omega_0^2$ are defined by $K_0/m$ and where it has been assumed that the

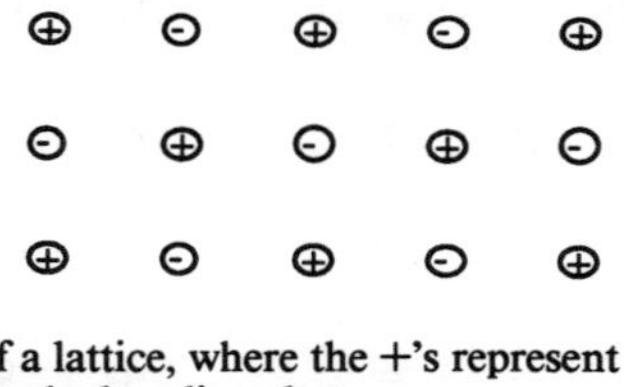

**FIGURE 6.8.**    A portion of a lattice, where the +'s represent lattice points with excess positive charge and the −'s are the bonding electrons.

frictional force $\mathbf{F}_f$ is given by

$$\mathbf{F}_f = my\dot{\mathbf{r}} \tag{6-5}$$

No real condensed materials can truly be isotropic. By analogy with Figure 6.8, Figure 6.9 depicts a portion of a lattice in a noncentrosymmetric crystal. Clearly the potential of (6-3) cannot describe the forces in this material. One could try to fix up the potential by adding a term in $x^3$ which would take into account the lack of symmetry of the potential in the $x$ direction. Taking

$$V(x,y,z) = \frac{1}{2} K_0 r^2 + \frac{1}{3} K_1 x^3 \tag{6-6}$$

leads to the system of equations

$$\ddot{x} + \gamma_x \dot{x} + \omega_{0_x}^2 x + \frac{K_1}{m} x^2 = F_{ext}|_x \tag{a}$$

$$\ddot{y} + \gamma_y \dot{y} + \omega_{0_y}^2 y = F_{ext}|_y \tag{b}$$

$$\ddot{z} + \gamma_z \dot{z} + \omega_{0_z}^2 z = F_{ext}|_z \tag{c} \tag{6-7}$$

There are various ways to interpret Equation (6-7a). Clearly, in Equation (6-4), if one were to assume a monochromatic driving force of the form

$$\mathbf{F}_{ext} = \mathbf{F} \cos \omega t \tag{6-8}$$

one could make the assumption that

$$\mathbf{r}(t) = \text{Re}[\tilde{\mathbf{r}}\, e^{-i\omega t}] \tag{6-9}$$

to find that

$$\tilde{\mathbf{r}} = \frac{\mathbf{F}}{\omega_0^2 - \omega^2 + i\omega\gamma} \tag{6-10}$$

This cannot be done with (6-7a), as the equation is nonlinear. However, if

**FIGURE 6.9.**    A portion of noncentrosymmetric crystal lattice.

(6-7) were driven with (6-8), it is easy to see that the $x$ equation would have a solution of the form

$$x(t) = \sum x_n^s \sin n\omega t + \sum x_m^c \cos n\omega t \tag{6-11}$$

The logic is that the squared term will couple a $\cos \omega t$ term with a D.C. term and a $\cos 2\omega t$ term. Having $\cos \omega t$ and $\cos 2\omega t$ terms in the squared term will generate $\cos 3\omega t$ terms, etc. The process described is known as *harmonic generation* and is related to the electrooptic effect.

Another way to look at Equation (6-7a) is to write it in the form

$$\ddot{x} + \gamma\dot{x} + \omega_{0x}^2(x)x = F_{ext}|_x \tag{6-12}$$

where the $\omega_{0x}^2(x)$ is given by

$$\omega_{0x}^2(x) = \omega_{0x}^2\left(1 + \frac{K_1}{m}x\right) \tag{6-13}$$

If the forcing term were given by

$$F_{ext} = F_0 + F_1 \cos \omega t \tag{6-14}$$

it becomes clear that, by changing the $F_0$ in (6-12), one can "tune" the resonant frequency $\omega_{0x}^2(x)$. This is basically the electrooptic effect.

The electrooptic tensor is defined by

$$\Delta\left(\frac{1}{n^2}\right) = r_{ij}E_j; \qquad i = 1,2,3,4,5,6; \qquad j = 1,2,3 \tag{6-15}$$

where the 1, 2, and 3 refer to crystallographic axes, which could be called $x$, $y$, and $z$, and the 4, 5, and 6 refer to axes which are located midway between $y$ and $z$, $x$ and $z$, and $x$ and $y$, respectively. Equation (6-15) corresponds to the fact that any product combination of $x$, $y$, and $z$ could have shown up in the potential expressions of (6-3) and (6-11). This means that static fields along any of the principal axes could have affected any of the resonant frequencies and therefore the indices along any of the principal axes or bisector directions. That there are six components to the $i$ index simply corresponds to the fact that symmetry requires that $n$, and therefore $\epsilon$ is a tensor of the form

$$\epsilon = \begin{bmatrix} \epsilon_{xx} & \epsilon_{xy} & \epsilon_{xz} \\ \epsilon_{yx} & \epsilon_{yy} & \epsilon_{yz} \\ \epsilon_{zx} & \epsilon_{zy} & \epsilon_{zz} \end{bmatrix} \tag{6-16}$$

## 6.3 ELECTROOPTIC DEVICES

The first electrooptic devices to be built were so-called bulk modulators. Say we have a material which has a nonzero electrooptic component $r_{33}$, as does LiNbO$_3$. We could put this component to work for us as illustrated in Figure 6.10. The idea is to put electrodes on the top and bottom of the crystal and then apply a voltage such that an electric field will be set up parallel to the crystallographic $z$ axis. In what follows, $z$ will generally be the direction of propagation, and therefore we will try to distinguish between crystallographic coordinates and optical coordinates. In Figure 6.10 the crystallographic coordinates are labelled 1, 2, and 3. The operation of the bulk phase shifter is as follows. With no voltage applied, the electric field would propagate through the crystal essentially as a plane wave in a homogeneous medium of index $n_0$. With voltage applied, there will be a change of index of

$$\Delta n_3 = - n_0^3 r_{33} E_3 \tag{6-17}$$

Therefore, in propagating from 0 to $z$ along the two crystallographic axes, the field will pick up a phase factor defined by

$$E_z(z) = e^{ik_0 n_0 z} \, e^{-ik_0 \int_0^z n_0^3 r_{33} E_3 \, dz} \, E_z(0) \tag{6-18}$$

where, for a uniform applied field $E_3$, the integral becomes trivial. The point of this is that the phase of the wave has been modified and therefore we have designed a phase modulator. This device has only two fatal flaws. One is that it is not a guided wave device and therefore will have to go into a bulky, complicated, high insertion loss system. The second flaw is related to the first but is not identical. Because the device is not guided wave, we are required to generate a uniform field throughout the crystal. For mechanical stability, the crystal probably has to be 1 mm thick. This means that our field, which is the voltage divided by the thickness, will be much weaker than if, for example,

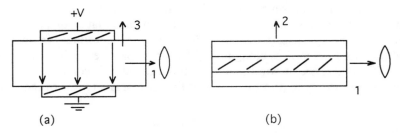

(a)                                        (b)

**FIGURE 6.10.**   Two views of bulk phase modulators: (a) from the excitation end and (b) from the top.

we were to drop all of the voltage over a single mode waveguide of width 5 $\mu$m. Because of this last effect, it turns out that the bulk modulators require voltages in the thousands of volts rather than in volts.

The question now might well arise as to how we could make an integrated optical, and therefore guided wave, device. Take for example the fabrication of a guided wave phase shifter in a LiNbO$_3$ substrate. The reason we might choose LiNbO$_3$ for this task is twofold. One is that the $r_{33}$ of LiNbO$_3$ is relatively large. That is,

$$r_{33}|_{\text{LiNbO}_3} = 30 \times 10^{-12} \frac{m}{v} \tag{6-19}$$

The second reason is that, despite its large $r_{33}$, LiNbO$_3$ has a relatively high Curie temperature of

$$T_C = 1250°C \tag{6-20}$$

This is very advantageous for processing, as we shall soon see.

Figure 6.11 illustrates the typical dimensions of a LiNbO$_3$ substrate with phase shifting guide. Obviously, the figure is not drawn to scale. The guide itself, to be single moded and fiber compatible, cannot be larger than 4–5 $\mu$m for operation at 0.85 $\mu$m. The gap between the electrodes will be of this same order of magnitude in order to generate a field structure as is illustrated in Figure 6.12. The important thing to note there is that the field is quite uniform and downward pointing in the guiding region. To get a feeling for the length of the device, let us calculate how far we would have to propagate the wave to get a $\pi$ phase shift where we assume that we want to operate with no more than 5 volts—that is, a digital circuit compatible voltage. The electric field will therefore be roughly 5 volts divided by 5 $\mu$m, or $10^6$ V/m. Writing that

$$\delta k = k_0 \Delta n L = -\pi \tag{6-21}$$

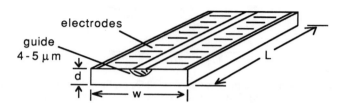

**FIGURE 6.11.**    Illustration (not to scale) of the sizes of importance for a LiNbO$_3$ waveguide device.

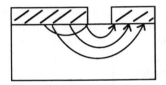

**FIGURE 6.12.**    Fields in a LiNbO₃ phase shifter.

gives us that

$$L = \frac{1}{n_0^3 r_{33} E} \frac{\lambda}{2} = 1600 \, \lambda \qquad (6\text{-}22)$$

which, for operation at 0.85 μm, would be a little more than a millimeter. This is indicative that integrated optic devices are long and thin.

To fabricate a device, we would want to borrow as much as possible from present-day techniques of planar circuit technology. The sequence of operations would appear as in Figure 6.13. The standard photolithographic techniques used for silicon processing are essentially the same ones used here. A difference is the very high aspect ratios used in the masking processes. De-

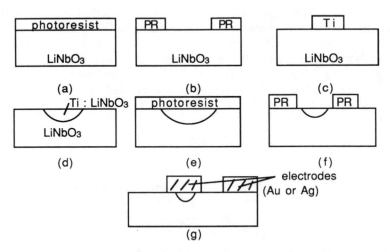

**FIGURE 6.13.**    Fabrication process for Ti : LiNbO₃ (a) First the substrate is coated with photoresist, which is then (b) photographically exposed where necessary and the excess removed such that (c) titanium can be evaporated on the surface for (d) subsequent indiffusion, and then the process of photoresist definition is again repeated (e-g) such that electrodes can be deposited.

vices can be centimeters long, yet channels are 2–3 $\mu$m and therefore require submicron alignment. Silicon circuits have a tendency to be more or less square. A second difference is the very high temperature at which the indiffusion must take place. It is generally done at more than 1000°C. This is a major reason why LiNbO$_3$ is the material of choice. Most materials with high electrooptic coefficients are somewhat unstable in the sense that their $T_C$'s are at best in the hundreds of degrees centigrade.

One notable result of the processing will be that the index in the channel will not be homogeneous. In fact, this is one of the big problems with automating integrated optics. Any computer-aided design (CAD) predictions require accurate knowledge of the index distribution. Yet this distribution is hard both to predict and measure due to the fact that the guides are single moded. Beyond this is the fact that we cannot use as simple an analysis as was used in (6-17) and (6-18) to find the total phase shift. We are forced to use something more complicated such as, for example, the approximate variational principle

$$\beta^2 = k_0^2 \frac{\int \psi^* n^2(x,y)\psi \, d^2x}{\int \psi^*\psi \, d^2x} \tag{6-23}$$

which can be derived from the wave equation

$$\frac{\partial^2}{\partial x^2}\psi + \frac{\partial^2}{\partial y^2}\psi + (k_0^2 n^2(x,y) - \beta^2)\psi = 0 \tag{6-24}$$

by multiplying through by $\psi^*$, integrating, and using the slowly varying approximation to rid oneself of the derivative terms. The idea behind (6-23) is that, if one has an idea of what $\psi$ looks like, one can get a first order approximation, which turns out to be actually second order in the error in the field, again, under the slowly varying approximation. At any rate, the variational principle predicts that

$$\delta\beta^2 = -k_0^2 r_{33}^2 \frac{\int n^2(x,y) \, \psi^* E_3^2 \psi \, d^2x}{\int \psi^*\psi \, d^2x} \tag{6-25}$$

and will be less than predicted by our earlier analysis. This is often taken into account by defining a fudge factor $\Gamma$, called the overlap, such that

$$\delta\beta = -k_0 r_{33} n^3 \, \Gamma E_z \tag{6-26}$$

A second device we want to consider here is the Mach-Zender interferometer. Top and end views of the device appear in Figure 6.14. To under-

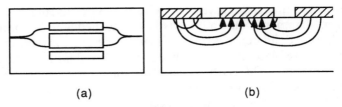

**FIGURE 6.14.**    Integrated Mach-Zender interferometer: (a) top view, (b) end view.

stand the device, one must think about the true modes of a two-waveguide structure. Consider the slab structure of Figure 6.15 and the sketches of its two lowest order modes, which we call $\psi$ symmetric, $\psi_s$, and $\psi$ antisymmetric, $\psi_a$. The variations of the propagation constants of these two modes are plotted as a function of the guide separation $d$ in Figure 6.16. When $d$ goes to zero, there is but a single mode guide, and therefore the antisymmetric mode cuts off. For $d$ becoming large, the modes become degenerate. This is because the guides decouple, and one could as well represent the fields by the individual guide fields $\psi_1$ and $\psi_2$, which in the large $d$ limit must be related to the $\psi_s$ and $\psi_a$ by

$$\psi_s = \frac{\psi_1 + \psi_2}{\sqrt{2}} \tag{a}$$

$$\psi_a = \frac{\psi_1 - \psi_2}{\sqrt{2}} \tag{b} \quad (6\text{-}27)$$

With the above considerations, we are ready to discuss the operation of the Mach-Zender interferometer. At the first junction, the field is split between the two channels, preferably equally. If the two channels are the same shape, this mode of the composite structure will be the even mode. The arms keep coming apart, however, until the guides are totally decoupled, and

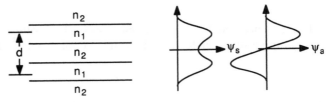

**FIGURE 6.15.**    Two-waveguide slab structure and the two lowest order modes that the structure could support.

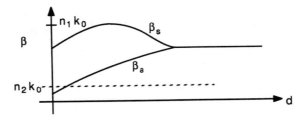

**FIGURE 6.16.**    Variation of the $\beta$s for the two lowest order modes of the structure of Figure 6.15 as a function of the guide separation $d$.

therefore the representation of (6-27) becomes applicable. With no voltage applied and for equal path length arms, the symmetric mode will be the one incident on the second $y$ junction. This mode would then be transmitted through the second junction, and 100% transmission through the device would be the result. If, however, a voltage is applied that is enough to produce a $\pi$ phase shift relative to the $\psi_1 \psi_2$ modes, then the antisymmetric mode will be incident on the second $y$ junction, and this mode will cut off and there will be zero transmission. The $T$ versus $V$ curve for this device is given in Figure 6.17. The curve is essentially a cosine squared curve. As is evident from the above discussion, this device can find application as a modulator and, in fact, a very fast one which can have bandwidths in the tens of gigahertz. If one makes one arm a little longer than the other in this device, one can effectively bias the device passively such that it operates about the $V\pi/2$ point of Figure 6.17. For small applied voltages about this point, the modulation will be linear in the voltage.

The next device to be discussed is the directional coupler of Figure 6.18. The device has various uses. Its fundamental operation is that, by applying a voltage, one changes the amount of light exiting from a channel. Its uses can

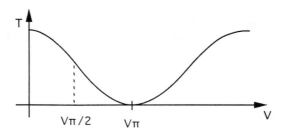

**FIGURE 6.17.**    Transmission versus voltage curve for the Mach-Zender device of Figure 6.14.

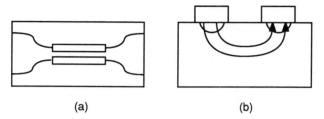

**FIGURE 6.18.** (a) Top and (b) end views of a directional coupler.

be summarized as in Figure 6.19, where the directional coupler is shown doing various operations.

As can be seen, the directional coupler is a complicated beast. It has straight sections at its end which must be bent together to form a coupling region and then pulled apart again to couple out. An important reason for these bent sections is that the channels can be no more than roughly one channel width apart in the coupling region, yet to be fiber compatible the channels must be minimally 125 $\mu$m (a fiber outer diameter) apart at input and output. As the analysis of the device proceeds, we shall see that there is another reason for pulling the guides apart.

For purposes of analysis, we break the directional coupler up into three sections as illustrated in Figure 6.20. We first carry out the analysis in regions III. We know that in region I we must consider the structure's modes to be made up of a symmetric mode $\psi_s$ and antisymmetric mode $\psi_a$. In region III, however, there modes will be degenerate and, further, they will be degenerate with the individual guide modes. Therefore, in regions III, relations (6-26) apply. In region I, we will assume that the disturbance in the guide, $\psi_g$, is expressible in the form

$$\psi_g = a_s(z)\psi_s e^{i\beta_s z} + a_a(z)\psi_a e^{i\beta_a z} \tag{6-28}$$

and therefore, to solve the directional coupler problem, we need to find $a_s(z_{II,I})$ and $a_a(z_{II,I})$. Finding the $a_s(a_{III,II})$ and $a_a(z_{III,II})$ is straightforward,

**FIGURE 6.19.** Three uses of directional couplers: (a) modulation, (b) switching and (c) mixing.

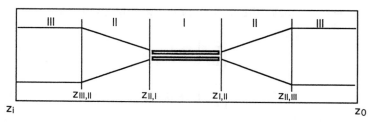

**FIGURE 6.20.**    Regions into which the directional coupler will be separated for purposes of analysis.

assuming that we can use overlap integrals. That is, if a disturbance $\psi_{in}$ is incident on the fiber endface, then we can write that

$$a_s(0) = \int \psi_{in} \frac{\psi_1 + \psi_2}{\sqrt{2}} d^2x \qquad \text{(a)}$$

$$a_a(0) = \int \psi_{in} \frac{\psi_1 - \psi_2}{\sqrt{2}} d^2x \qquad \text{(b)} \quad (6\text{-}29)$$

As the guides are uncoupled, we can further write

$$a_s(z_{III,II}) = a_s(0) \qquad \text{(a)}$$

$$a_a(z_{III,II}) = a_a(0) \qquad \text{(b)} \quad (6\text{-}30)$$

One would expect that the next thing to do would be to solve the problem in the bent waveguide regions II and thereby find $a_s(z_{II,I})$ and $a_a(z_{II,I})$. Unfortunately, this problem has not been solved. In a later paragraph, some discussion will be given to how to solve this problem, but for now we shall ignore this region and simply say that $a_s(z_{II,I})$ and $a_a(z_{II,I})$ are identical with their values at $z = z_{III,II}$. This cannot be an especially good approximation.

To solve the problem in region I, we say that the $E_x$ field looks like the $\psi_g$ in Equation (6-28). To find coupled equations, we need to know the corresponding $H$ field, which we can find from Maxwell's curl equation to be

$$H_y = \frac{1}{i\omega\mu_0} \frac{\partial E_x}{\partial z} \qquad (6\text{-}31)$$

It is clear that there will be no modes at all if the $a$'s vary too rapidly. We therefore want to make the slowly varying approximation that

$$\left|\frac{da_s}{dz}\right| \ll |\beta_s a_s| \qquad \text{(a)}$$

$$\left|\frac{da_a}{dz}\right| \ll |\beta_a a_a| \qquad \text{(b)} \quad \text{(6-32)}$$

to find that (6-31) expands to

$$H_y = \frac{a_s}{\eta_s} \psi_s e^{i\beta_s z} + \frac{a_a}{\eta_a} \psi_a e^{i\beta_a z} \qquad \text{(6-33)}$$

which is a most reassuring result, as this is indeed the $H_y$ we would expect to go with an $E_x$ of the form of (6-28). Using Maxwell's other curl equation

$$\nabla \times \mathbf{H} = -i\omega(\varepsilon + \Delta\varepsilon)\mathbf{E} \qquad \text{(6-34)}$$

one finds without any further approximation the coupled equations

$$\frac{\partial a_s}{\partial z} = i\chi e^{-i\Delta\beta z} \qquad \text{(a)}$$

$$\frac{\partial a_a}{\partial z} = i\chi e^{i\Delta\beta z} \qquad \text{(b)} \quad \text{(6-35)}$$

where $\chi$ and $\Delta\beta$ are given by

$$\chi = \frac{1}{2} \int \psi_s \Delta\varepsilon \psi_a \, d^2x \qquad \text{(a)}$$

$$\Delta\beta = \beta_s - \beta_a \qquad \text{(b)} \quad \text{(6-36)}$$

The coupled equations of (6-35) can be solved in straight waveguide region of I to yield

$$a_s(z) = e^{-i\frac{\Delta\beta}{2}z} \left[ a_s(z_{\text{II,I}}) \cos bz + i\left(\frac{\Delta\beta}{2b}a_s(z_{\text{II,I}}) + \frac{\chi}{b}a_a(z_{\text{II,I}})\right) \sin bz \right] \text{(a)}$$

$$a_a(z) = e^{+i\frac{\Delta\beta}{2}z} \left[ a_a(z_{\text{II,I}}) \cos bz - i\left(\frac{\Delta\beta}{2b}a_a(z_{\text{II,I}}) - \frac{\chi}{b}a_s(z_{\text{II,I}})\right) \sin bz \right] \text{(b) (6-37)}$$

where $b$ is given by

$$b = \sqrt{\frac{\Delta\beta^2}{4} + \chi^2}$$

(6-38)

We now wish to look at a couple of limits of these solutions. Let us say, for example, that the voltage is turned off and therefore $\chi \to 0$. Then one has the solution

$$a_s(z) = a_s(z_{\mathrm{II,I}})$$

(a)

$$a_a(z) = a_a(z_{\mathrm{II,I}})$$

(b)   (6-39)

We do not know what happens between $z_{\mathrm{I,II}}$ and $z_{\mathrm{II,III}}$, so we shall just set $a_s$ and $a_a$ at $z_{\mathrm{II,III}}$ equal to $a_s$ and $a_a$ at $z_{\mathrm{I,II}}$. We then can use Equations (6-37) and

$$a_s(z_{\mathrm{II,III}}) = \frac{a_1(z_{\mathrm{II,III}}) + a_2(z_{\mathrm{II,III}})}{\sqrt{2}}$$

(a)

$$a_a(z_{\mathrm{II,III}}) = \frac{a_1(z_{\mathrm{II,III}}) - a_2(z_{\mathrm{II,III}})}{\sqrt{2}}$$

(b)   (6-40)

to show that the $E_x$ of (6-28) can be expressed in region III as

$$\psi_g = a_1(\psi_1 \cos \Delta\beta L - i\psi_2 \sin \Delta\beta L) + a_2(\psi_2 \cos \Delta\beta L - i\psi_1 \sin \Delta\beta L) \quad (6\text{-}41)$$

where $L$ is the length of the coupling region. It is not quite clear what to use for this length due to the unsolved problem in regions II. If we were to take the input as being $\psi_{\mathrm{in}} = \psi_1$, then we see that the power seems to oscillate back and forth between the guides as a function of the length of the guiding region as is depicted in Figure 6.21. The power is actually oscillating back and forth between the guides, but what is happening is that in the coupling region the

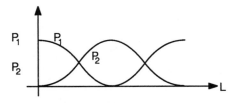

**FIGURE 6.21.**    Variation of the power coming out of, respectively, guides 1 or 2 as a function of the length of the coupling region.

interference between the symmetric and antisymmetric modes is forming an oscillating dipole pattern in propagation distance.

Let us now see what happens when a large voltage is applied, or when $\chi \gg \Delta\beta$. Then the solution reduces to

$$a_s(z_{\text{II,III}}) = a_s(z_{\text{III,II}}) \cos \chi L + a_a(z_{\text{III,II}}) \sin \chi L \qquad \text{(a)}$$

$$a_a(z_{\text{II,III}}) = a_a(z_{\text{III,II}}) \cos \chi L + a_s(z_{\text{III,II}}) \sin \chi L \qquad \text{(b)} \quad \text{(6-42)}$$

As in the last paragraph, we can now express the $E_x$ in region III in terms of the individual guide quantities to find

$$\psi_g = a_1 e^{i\chi L}\psi_1 + a_2 e^{-i\chi L}\psi_2 \qquad (6\text{-}43)$$

which shows explicitly that no coupling has taken place. The large voltage has so distorted the index profile that the $\psi_s$ and $\psi_a$ modes are no longer the modes, but the $\psi_1$ and $\psi_2$ are approximately. Our knowledge of the directional coupler problem can be summed up by the plot of $P_2$ versus voltage, as is depicted in Figure 6.22. The basic design strategy is to try to make the device length equal to exactly one coupling length $l_c = \pi/\Delta\beta$ so that when power is coupled into the 1 channel, it will come out in the 2 channel. This is rather hard to do, as one does not know $\Delta\beta$ to a very high accuracy, and further we do not really know the $L$ of the interaction length because of the problem in regions II. This is the reason that the maximum of $P_2$ is not quite at zero voltage. The first two minima of the curve are going to lie at the so-called switching voltage, as this is the voltage that has to be applied to "switch" the power from $P_2$ to $P_1$. These minima will not quite be equal to zero. For higher and higher voltages, the interference effect washes out and the power will all come out in guide 1.

In the literature, there is a body of material which expands in terms of the individual guide modes even in the coupling region. In this paragraph, we wish to see what approximation this actually entails. Figure 6.23 can help us visualize what the meaning of some of the approximations are. The idea is

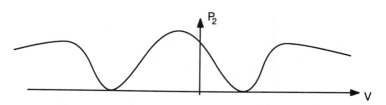

**FIGURE 6.22.**    Possible $P_2$-$V$ characteristic of a directional coupler.

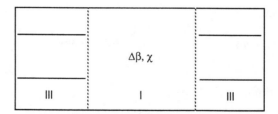

**FIGURE 6.23.**    The problems associated with individual guide mode coupled modes.

that we completely ignore the taper regions and consider region I to be a magical region in which the guides are uncoupled but a $\Delta\beta$ and $\chi$ mathematically appear. Mathematically, this picture does have some validity, as can be seem from the following derivation. Let us define rapidly varying coefficients by

$$\tilde{a}_s = a_s e^{i\beta_s z} \qquad \text{(a)}$$

$$\tilde{a}_a = a_a e^{i\beta_a z} \qquad \text{(b)} \quad \text{(6-44)}$$

to transform the coupled equations of (6-35) into the form

$$\frac{\partial \tilde{a}_s}{\partial z} + i\beta_s \tilde{a}_s = i\chi \tilde{a}_a \qquad \text{(a)}$$

$$\frac{\partial \tilde{a}_a}{\partial z} + i\beta_a \tilde{a}_a = i\chi \tilde{a}_s \qquad \text{(b)} \quad \text{(6-45)}$$

We can then define

$$\tilde{a}_1 = \frac{\tilde{a}_s + \tilde{a}_a}{\sqrt{2}} \qquad \text{(a)}$$

$$\tilde{a}_2 = \frac{\tilde{a}_s - \tilde{a}_a}{\sqrt{2}} \qquad \text{(b)} \quad \text{(6-46)}$$

to obtain (6-45) in the form

$$\frac{\partial \tilde{a}_1}{\partial z} + i(\bar{\beta} - \chi)\tilde{a}_1 = i\Delta\beta \tilde{a}_2 \qquad \text{(a)}$$

$$\frac{\partial \tilde{a}_2}{\partial z} + i(\bar{\beta} - \chi)\tilde{a}_2 = i\Delta\beta \tilde{a}_1 \qquad \text{(b)} \quad \text{(6-47)}$$

where $\beta$ is defined by

$$\bar{\beta} = \frac{\beta_1 + \beta_2}{2} \tag{6-48}$$

Making definitions

$$\tilde{a}_1 = a_1 e^{i\bar{\beta}z} e^{-i\chi} \tag{a}$$

$$\tilde{a}_2 = a_2 e^{i\bar{\beta}z} e^{i\chi} \tag{b} \quad (6\text{-}49)$$

one finds that

$$\frac{\partial a_1}{\partial z} = \frac{i\Delta\beta}{2} a_2 e^{2i\chi z} \tag{a}$$

$$\frac{\partial a_2}{\partial z} = \frac{i\Delta\beta}{2} a_1 e^{-2i\chi z} \tag{b} \quad (6\text{-}50)$$

and lo and behold, we have gotten back the coupled mode equations with $\chi$ and $\Delta\beta/2$ interchanged. Much has gone into interpreting these equations. Unfortunately, though, they are not terribly internally consistent. One cannot realistically use the $\psi_1$ and $\psi_2$ functions to calculate $\Delta\beta$ and $\chi$, as these values are only defined for the actual modes $\psi_s$ and $\psi_a$. Further, one will never be able to solve the bend problem with these modes.

We now wish to see how we might solve the bend problem. The idea we have always used in coupled modes is to expand about the solution to a problem that we know how to solve. Clearly, we can solve the problem of two straight waveguides at a distance of $d(z)$ apart. Generally what we do is, from the outset, to write

$$\psi(x,y,z,t) = \text{Re}[\psi(x,y)e^{i\beta z}e^{-i\omega t}] \tag{6-51}$$

But we cannot do that here, as there is no $\beta$. But, in the slowly varying spirit, we could try to take

$$\psi(x,y,z,t) = \text{Re}[\psi_z(x,y)e^{i\int\beta(z)\,dz}e^{-i\omega t}] \tag{6-52}$$

where the $\psi_z(x,y)$ and $\beta(z)$ are the solutions to the problem of two waveguides separated by a distance $d(z)$. The problem that remains is to try to find a way to preserve the modal characteristics of the propagation while finding out about the coupling.

Previously, we used the curl **E** equation to define the modes and the curl **H** equation to get the coupling. As we have no $\Delta\epsilon$ in our problem now, there is no use in following this same approach. Therefore, let us say that

$$H_y = \frac{a_s(z)\psi_{sz}(x,y)}{\eta_s(z)} e^{i\int\beta_s(z)\,dz} + \frac{a_a(z)\psi_{az}(x,y)}{\eta_a(z)} e^{i\int\beta_a(z)\,dz} \qquad (6\text{-}53)$$

and use curl **H** to find

$$E_x = \frac{\epsilon_s(z)}{\epsilon(x,y,z)}\,\psi_{sz}(x,y)e^{i\int\beta_s(z)\,dz} + \frac{\epsilon_a(z)}{\epsilon(x,y,z)}\,\psi_{az}(x,y)e^{i\int\beta_a(z)\,dz} \qquad (6\text{-}54)$$

where a slowly varying approximation has been made. Indeed, (6-53) and (6-54) look like the relations a mode should satisfy, so we seem to be going in the right direction. By now using curl **E**, it is straightforward to show

$$\frac{\partial a_x}{\partial z} = i\chi_{ss}(z)a_s(z) + i\chi_{sa}(z)a_a(z)e^{-i\int\Delta\beta\,dz} \qquad \text{(a)}$$

$$\frac{\partial a_a}{\partial z} = i\chi_{aa}(z)a_a(z) + i\chi_{as}(z)a_s(z)e^{i\int\Delta\beta\,dz} \qquad \text{(b)} \quad (6\text{-}55)$$

where the $\chi$'s are given by

$$\chi_{ss}(z) = \frac{1}{2}\int \frac{\partial}{\partial z}\left[\psi_{sz}^2(x,y)\frac{\epsilon_s(z)}{\epsilon(x,y,z)}\right]d^2x \qquad \text{(a)}$$

$$\chi_{sa}(z) = \frac{1}{2}\int \frac{\partial}{\partial z}\left[\psi_{sz}(x,y)\frac{\epsilon_s(z)}{\epsilon(x,y,z)}\psi_{az}(x,y)\right]d^2x \qquad \text{(b)}$$

$$\chi_{as}(z) = \frac{1}{2}\int \frac{\partial}{\partial z}\left[\psi_{sz}(x,y)\frac{\epsilon_a(z)}{\epsilon(x,y,z)}\psi_{az}(x,y)\right]d^2x \qquad \text{(c)}$$

$$\chi_{aa}(z) = \frac{1}{2}\int \frac{\partial}{\partial z}\left[\psi_{az}^2(x,y)\frac{\epsilon_a(z)}{\epsilon(x,y,z)}\right]d^2x \qquad \text{(d)} \quad (6\text{-}56)$$

Note that for purely symmetric taper regions, $\epsilon(x,y,z)$ is an even function of $x$ and $y$ and therefore one finds that

$$\chi_{sa} = \chi_{as} = 0 \qquad (6\text{-}57)$$

Using this fact in (6-55) leads to the conclusion that the taper will induce a pure phase shift on each of the mode coefficients. In the general case of non-symmetric taper, however, there will be mode coupling as well as phase shift.

The last device we wish to consider is the polarization rotator. As fibers are not generally polarization preserving, it can be very useful to have a device which can take an arbitrary polarization state to another arbitrary polarization state. We shall presently see how that can be done.

To carry out polarization rotation, it is necessary to use another element of the electrooptic tensor. LiNbO$_3$ has an $r_{51}$ element which is almost as large as $r_{33}$. A device which uses this coefficient might be designed as the device sketched in Figure 6.24. As usual, we take the propagation direction to be $z$, despite the fact that this does not agree with the crystallographic axes. With this, we find that the transverse $\epsilon$ tensor will become

$$\epsilon_t = \begin{bmatrix} \epsilon_o & \epsilon_{xy} \\ \epsilon_{xy} & \epsilon_e \end{bmatrix} = \begin{bmatrix} \epsilon_o & 0 \\ 0 & \epsilon_e \end{bmatrix} + \begin{bmatrix} 0 & \epsilon_{xy} \\ \epsilon_{xy} & 0 \end{bmatrix} \tag{6-58}$$

where the $o$'s and $e$'s in the subscripts stand for ordinary and extraordinary. Say the transverse electric field is given by

$$\mathbf{E}_t = a_x \psi_x e^{i\beta_x z} \hat{e}_x + a_y \psi_y e^{i\beta_y z} \hat{e}_y \tag{6-59}$$

Using the slowly varying approximation in curl E gives

$$\mathbf{H}_t = \frac{a_x \psi_x}{\eta_x} e^{i\beta_x z} \hat{e}_x - \frac{a_y \psi_y}{\eta_y} e^{i\beta_y z} \hat{e}_y \tag{6-60}$$

which is indeed what one would have hoped for. Plugging into curl $\mathbf{H}$, one finds without further ado that

$$\frac{\partial a_x}{\partial z} = i\chi e^{-i\Delta\beta z} a_y \tag{a}$$

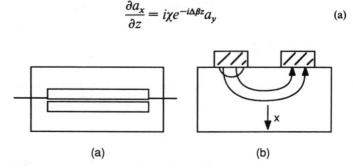

(a)                                          (b)

**FIGURE 6.24.**    (a) Top view and (b) end view of a device which uses $r_{51}$.

$$\frac{\partial a_y}{\partial z} = i\chi e^{i\Delta\beta z}a_x \qquad \text{(b)} \quad (6\text{-}61)$$

which have the solution as in (6-37).

Let us consider the limit where $\Delta\beta \to 0$. In this limit, we see that the $a_x$ and $a_y$ at a plane $z_2$ at the output of the device will be related to the $a_x$ and $a_y$ at a plane $z_1$ at the input of the device by

$$\begin{bmatrix} a_x(z_2) \\ a_y(z_2) \end{bmatrix} = \begin{bmatrix} \cos\theta & i\sin\theta \\ i\sin\theta & \cos\theta \end{bmatrix} \begin{bmatrix} a_x(z_1) \\ a_y(z_1) \end{bmatrix} \qquad (6\text{-}62)$$

where $\theta = \chi(z_2 - z_1)$. For linear $x$ or $y$ states, this looks something like a rotation. However, for example, for a 45° polarized wave incident, nothing happens. If we put in a general state of the form

$$\begin{bmatrix} a_x(z_1) \\ a_y(z_2) \end{bmatrix} = \frac{1}{\sqrt{c_x^2 + c_y^2}} \begin{bmatrix} c_x \\ c_y e^{i\gamma} \end{bmatrix} \qquad (6\text{-}63)$$

we get an extremely complicated field out. What we really want is a matrix of the form

$$M_\theta = \begin{bmatrix} \cos\theta & -\sin\theta \\ \sin\theta & \cos\theta \end{bmatrix} \qquad (6\text{-}64)$$

But this only differs in phase from the matrix in (6-62). We could try to fix this up by putting phase shifters before and after the $r_{51}$ element, as depicted in Figure 6.25. If we call the polarization vector at the coordinate $z_i$, $\mathbf{a}_i$, then we can write that

$$\mathbf{a}_3 = M_{23}\mathbf{a}_2 = M_{23}M_{12}\mathbf{a}_1 = M_{23}M_{12}M_{01}\mathbf{a}_0 \qquad (6\text{-}65)$$

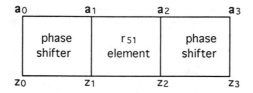

**FIGURE 6.25.**    Composite rotator.

We have $M_{12}$. $M_{23}$ and $M_{01}$ can be taken to be of the forms

$$M_{01} = \begin{bmatrix} 1 & 0 \\ 0 & e^{\delta_{01}} \end{bmatrix} \qquad \text{(a)}$$

$$M_{23} = \begin{bmatrix} 1 & 0 \\ 0 & e^{\delta_{23}} \end{bmatrix} \qquad \text{(b)} \quad \text{(6-66)}$$

to obtain for the total $M$

$$M = \begin{bmatrix} \cos\theta & ie^{i\delta_{01}}\sin\theta \\ ie^{i\delta_{23}}\sin\theta & e^{i(\delta_{23}+\delta_{01})}\cos\theta \end{bmatrix} \qquad \text{(6-67)}$$

Clearly, by taking $\delta_{01} = \pi/2$ and $\delta_{23} = -\pi/2$, the result becomes (6-64). An added advantage of this solution is that the $\delta_{01}$ can be adjusted to take out any unwanted phase from the incident distribution and the $\delta_{23}$ modified to give any arbitrary state at the output.

We have now shown ourselves that we can make a rotator device if we

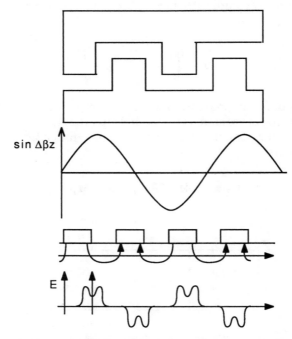

**FIGURE 6.26.**     Interdigitated electrode structure and its field structure relative to a plot of $\sin\Delta\beta z$.

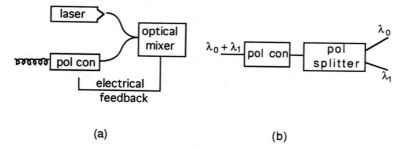

**FIGURE 6.27.**    Sketches of two applications of polarization controllers in (a) a hetero-dyne optical receiver and (b) a wavelength demultiplexer.

have $\Delta\beta = 0$. Unfortunately, $\Delta\beta \neq 0$ and is not even close. In general, the transverse field will be of the form

$$\mathbf{E}_t = e^{i\beta_x z} [c_x \hat{e}_x + c_y e^{-i\Delta\beta z} \hat{e}_y] \qquad (6\text{-}68)$$

Only near points where $\Delta\beta = (2n + 1)\pi$ will we be able to rotate the polarization. This can be noticed by applying $M_{12}$ to a state with the $y$ coefficient $i$ out of phase with the $x$ coefficient. The result is

$$\begin{bmatrix} \cos\theta & i\sin\theta \\ i\sin\theta & \cos\theta \end{bmatrix} \begin{bmatrix} c_x \\ ic_y \end{bmatrix} = \begin{bmatrix} c_x\sin\theta - c_y\cos\theta \\ i(c_x\sin\theta + c_y\cos\theta) \end{bmatrix} = \begin{bmatrix} c'_x \\ ic'_y \end{bmatrix} \qquad (6\text{-}69)$$

where the output is simply a rotated version of $[c_x, c_y]^\mathrm{T}$, but again with $i$ out of phase. This is to say that the $r_{51}$ section only performs a pure rotation on a special state that must be specially prepared. Figure 6.26 depicts a possible solution to this. The idea is to make an electrode configuration that only applies a field when the polarizations are in the proper phase to be rotated. Indeed, such devices are workable and can have such applications as are sketched in Figure 6.27.

## PROBLEMS

1. The "spring" equations describing a simple electrooptic medium might be written in the form

$$\ddot{x} + \gamma_x \dot{x} + \omega_x^2 x + \Delta\omega_{xy}^2 y = -\frac{eE_x}{m_x}$$

$$\ddot{y} + \gamma_y \dot{y} + \omega_y^2 y + \Delta\omega_{yx}^2 x = -\frac{eE_y}{m_y}$$

(a) Try to solve these equations for a harmonic wave of arbitrary polarization incident in the $z$ direction.

(b) Say that a static field of strength $E_{y0}$ is applied to a material whose polarizability is described by the differential system above. Find an expression for the change $\Delta n_x$ in the index of refraction of the $x$ polarization state due to this perturbation.

2. Here we wish to consider the effects of some perturbations on the operation of a Mach-Zender interferometer. A perfect one would have symmetric arms of equal length and therefore would have 100% transmission with zero voltage and 0% transmission with $V = V_\pi$. What happens to this scenario if:

(a) One arm has a wider waveguide than the other?

(b) The first $y$-junction does not split 50% to 50%?

(c) One arm is longer than the other?

(d) How would one purposefully mismatch the arms so that the interferometer has maximum sensitivity to voltages $V \ll V_\pi$?

3. We wish to sketch the $P$ versus $V$ curves of Mach-Zender interferometers for operating wavelengths $0.5\lambda_0$, $\lambda_0$, and $2\lambda_0$. The Mach-Zenders are

(a) Equal path lengths;

(b) Passive bias for $\lambda_0$, such that 1/2 power is transmitted for zero bias voltage at $\lambda_0$.

(c) Zero transmission for zero bias at $\lambda_0$.

4. Repeatedly in the chapter, we derived coupled mode equations from the curl equations by assuming that a perturbation $\Delta\epsilon$ was small enough that the modal coefficients $c_1$ and $c_2$ in the expression

$$\psi(x,z) = c_1(z)u_1(x)e^{i\beta_1 z} + c_2(z)u_2(x)e^{i\beta_2 z}$$

were slowly varying, then using the curl E equation to find the (roughly) modal form of $H$, and then using this form in curl $H$ to obtain coupled mode equations. Here, you should repeat such a calculation but for a magnetic material with $\epsilon = \epsilon_0$ and $\mu = \mu_g + \Delta\mu$ by assuming modes $\psi_1$ and $\psi_2$ with coefficients $c_1$ and $c_2$, using curl $H$ to obtain $E$ and curl $E$ to obtain coupled equations.

5. Consider the following system of equations:

$$\frac{\partial c_1}{\partial t} = M_{12}c_2$$

$$\frac{\partial c_2}{\partial t} = M_{21}c_1$$

(a) Obtain two decoupled second-order equations from this system.
(b) Express the general form of solution to the equations of part (a).
(c) Find the conditions on $M_{12}$ and $M_{21}$ under which the quantity $c_1^2 + c_2^2$ is conserved, i.e.,

$$\frac{\partial}{\partial t} \left( |c_1|^2 + |c_2|^2 \right) = 0$$

6. Say we have the following system of equations:

$$\frac{d\tilde{a}_1}{dz} + i\beta_1 \tilde{a}_1 = ix_{12}\tilde{a}_2$$

$$\frac{d\tilde{a}_2}{dz} + i\beta_2 \tilde{a}_2 = ix_{21}\tilde{a}_2$$

Transform these equations to the "usual" coupled mode form, in which a derivative of one coefficient is equal to a coefficient times the other coefficient.

7. The transfer function of a directional coupler can be written in the form

$$\begin{bmatrix} a_s(z) \\ a_a(z) \end{bmatrix} = M \begin{bmatrix} a_s(0) \\ a_a(0) \end{bmatrix}$$

(a) Say we wish to perform the operation

$$M_\theta = \begin{bmatrix} \cos\theta & \sin\theta \\ -\sin\theta & \cos\theta \end{bmatrix}$$

on an incoming vector $[a_1, a_2]^T$ where, for separated guides

$$a_1 = \frac{a_s + a_a}{\sqrt{2}}$$

$$a_2 = \frac{a_s - a_a}{\sqrt{2}}$$

are the individual guide modes. How close can the directional coupler come to performing this operation?

(b) The arms of a directional coupler must be brought together. Say that the phases of the symmetric and antisymmetric amplitudes get shuffled in this transition region, such that one must write

$$\begin{bmatrix} a_s(z) \\ a_a(z) \end{bmatrix} = M \begin{bmatrix} a_s(0) \\ a_a(0)e^{i\phi} \end{bmatrix}$$

What effect will this have on $M_\theta$? How could one try to correct this?

8. This problem concerns trying to determine the error involved in using Equation (6-50) instead of Equation (6-45) in determining the propagation within a coupled waveguide section. It also requires using a computer program to solve for the propagation constants of a channel using, for example, the effective index method.

(a) Pick a coupled waveguide geometry such that one finds a $\Delta\beta = \beta_s - \beta_a$ such that the coupling length $\ell_c = \pi/\Delta\beta$ is less than a couple thousand wavelengths.

(b) Calculate the $\psi_s, \psi_a, \psi_1$ and $\psi_2$ for the above configuration and plot. For parts (c) and (d) you will need to use perturbation theory if your program assumes a symmetric index profile.

(c) Pick a perfect antisymmetric perturbation of some value $\Delta n$ and calculate the $\psi$ calculated from using $\psi_1$ and $\psi_2$ versus that calculated using $\psi_s$ and $\psi_a$.

(d) Calculate the actual first two $\psi$'s of the waveguide if the $\Delta n$ were included into the waveguide structure. How different are they from $\psi_s$ and $\psi_a$? From $\psi_1$ and $\psi_2$?

9. Say one has found the modes of a slab geometry structure to consist of $N$ TE modes $\psi_i$ and $N$ TM modes $\phi_i$ such that any $E_y$ in the guide can be expressed in the form

$$E_y = \sum a_n \psi_n(x)$$

and $H_y$ by

$$H_y = \sum b_n \phi_n(x)$$

Given that the guide is perturbed by an index distribution $\Delta n(x,z)$, find:

(a) The coupling equation for the $a_n$'s;

(b) The coupling equation for the $b_n$'s.

(c) What if $\Delta n(x,z) = $ constant? What would this correspond to in a quantum mechanical system?

(d) What happens if $\Delta n$ is a function of time, but not of space?

10. Let us consider the propagation of a $z$-directed plane wave in a periodically striated medium whose index distribution is given by

$$n^2(z) = n_1^2 + \Delta n^2 \cos Kz$$

Flocquet proved that the solution to a differential equation with periodic coefficients could be written in the form

$$\psi(z) = e^{i\beta z} \sum_{n \to -\infty}^{\infty} a_n e^{inKz}$$

(a) Write down the equation satisfied by a transverse component $\psi$ of the electric field in the above defined periodic medium.

(b) Find a matrix equation for the determination of $\beta$ and the $a_N$'s of Flocquet's solution.

(c) Often times, when $\Delta n^2 \ll n_1^2$, one approximates $\beta \approx k_0 n_1$ and that $a_n = a_n(z)$ is a very slowly varying function of $z$. Make these ap-

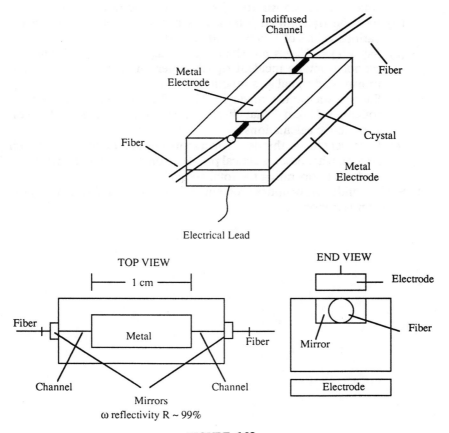

**FIGURE 6.28.**

proximations and derive the system of first-order differential equations satisified by the $a_n(z)$.

(d) Truncate the equations of (c) to include only $a_0(z)$ and $a_{-1}(z)$ and then solve the resulting equations. Plot the propagation constant of $a_0(z)$ as a function of $\beta$, the wave vector.

11. Consider the device depicted in Figure 6.28. The crystal which is host to the weakly guiding channel waveguide is electrooptic (i.e., whose index of refraction can be varied by applying a field between the electrodes) with an electrooptic coefficient of $r = \dfrac{10^{-5}}{V}$ (i.e., $\Delta n = r\, V_{applied}$) where $n_0 \approx 2$. Assume the electrodes are both close to 1 cm long, and in (a)–(c), that the waveguide is lossless.

(a) If the incident signal is broadband noise centered on 0.85 $\mu$m, what signal is transmitted for the applied voltage $V = 0$?

(b) Plot the transmitted intensity as a function of wavelength for various values of $V$ between 0 and 10 volts.

(c) Say $V(t) = V_0 \cos \omega t$ where $V_0$ is roughly 10 volts. For almost monochromatic incident signal center on 0.85 $\mu$m, what is the transmitted signal as a function of time?

(d) Such a waveguide made in Ti:LiNbO$_3$ would have losses of roughly 1 dB/cm. What effect would inclusion of this loss have on your above calculations?

(e) If we were to use this device in a semiconductor laser diode driven system, what other practical problem (on top of that in (d)) would we have in operating the total system?

12. Solve a guided wave optics problem of your choosing. Explain why the problem is important.

# List of Symbols

## LATIN SYMBOLS

$a$      $a_j$ mode coefficient, $a$ radius
$A$      $A$ area, d$\mathbf{A}$ element of area
$b$      $b$ photon amplitude
$c$      $c$ speed of light, $c_i$ amplitude coefficient
$C$      $C$ arbitrary constant
$d$      $d_\mu$ microscopic inversion parameter
$D$      $\mathbf{D}$ displacement vector, $D$ inversion parameter
$e$      $-e$ charge of electron, $\hat{\mathbf{e}}_i$ unit vector in $i$ direction
$E$      $\mathbf{E}$ electric field vector, $E_i$ energy level, $E$ energy
$f$      $f$ frequency
$g$      $g_{\mu\lambda}$ atomic field coupling parameter
$h$      $h$ Planck's constant
$\hbar$      $\hbar$ Planck's reduced constant $= h/2\pi$
$H$      $\mathbf{H}$ magnetic field vector, $H$ Hamiltonian
$i$      $i$ an alternating current
$I$      $I$ intensity, $I$ a direct current
$J$      $\mathbf{J}$ current density
$k$      $k$ plane wave propagation constant, $k_B$ Boltzmann's constant
$K$      $K$ an arbitrary constant
$l$      $l$ modal azimuthal subscript
$L$      $L$ a length, $\mathbf{L}$ left matrix
$m$      $m$ mass
$M$      $\mathbf{M}$ characteristic matrix
$n$      $n$ index of refraction, $n$ number density
$N$      $N$ normalization factor
$p$      $\mathbf{p}$ microscopic polarizability, $\hat{\mathbf{p}}$ dipole operator

| | |
|---|---|
| $P$ | $P$ power, $\mathbf{P}$ polarization vector, $\mathscr{P}$ intrinsic dipole moment |
| $q$ | $q$ a charge |
| $r$ | $r$ reflection coefficient, $\mathbf{r}$ radius velocity |
| $R$ | $R$ reflectivity, $\boldsymbol{R}$ right matrix |
| $S$ | $\mathbf{S}$ Poynting vector |
| $t$ | $t$ transmission coefficient, $t$ time |
| $T$ | $T$ transmissivity, $\boldsymbol{T}$ transmission matrix, $T$ temperature |
| $u$ | $u_i$ time independent wave function |
| $v$ | $v$ velocity of propagation |
| $V$ | $\mathbf{V}$ vector, $V$ guide, $V$ number, $V$ volume, $V$ potential function, $V_{ij}$ coupling coefficients |
| $w$ | $w$ a rate |
| $W$ | $W$ spectral density |
| $x$ | $x$ coordinate |
| $X$ | $X$ crystallographic axis |
| $y$ | $y$ coordinate |
| $Y$ | $Y$ crystallographic axis |
| $z$ | $z$ coordinate |
| $Z$ | $Z$ crystallographic axis |

## GREEK AND SPECIAL SYMBOLS

| | |
|---|---|
| $\alpha$ | $\alpha$ microscopic polarizability |
| $\beta$ | $\beta$ propagation constant |
| $\gamma$ | $\gamma$ transverse propagation constant, $\gamma$ atomic damping constant |
| $\delta$ | $\delta$ phase angle |
| $\epsilon$ | $\epsilon$ permittivity, $\epsilon_g$ energy |
| $\eta$ | $\eta$ impedance |
| $\theta$ | $\theta$ an angle |
| $\kappa$ | $\kappa$ transverse decay constant |
| $\lambda$ | $\lambda$ wavelength, $\lambda$ eigenvalue |
| $\mu$ | $\mu$ permeability |
| $\nabla$ | $\nabla$ gradient operator |
| $\nabla^2$ | $\nabla^2$ Laplacian operator |
| $\nu$ | $\nu$ a radial modal subscript |
| $\rho$ | $\rho$ charge density |
| $\sigma$ | $\sigma$ conductivity |
| $\tau$ | $\tau$ decay constant |
| $\phi$ | $\phi$ phase angle |
| $\psi$ | $\psi$ wave function, $\boldsymbol{\psi}$ vector wave function |
| $\omega$ | $\omega$ angular frequency |
| $\Omega$ | $\Omega$ solid angle |

# Index

Adiabatic elimination of variables, 72
Alpha index profile, 154
Ancient Greek optics, 3-4
Archimedes burning glass, 3-4
Asymmetric slab
  eigenvalue equation, 22
  field solutions, 20-21
Atomic dipole moment, 35
Azimuthal mode number
  defined for LP modes, 158
  defined in terms of ray coordinates, 187

Bragg condition, 113
Bragg effect, 113-114
Background index, 109-110
Backscattering
  measurements, 167, 208, 219-220
  optical time domain reflectometry
    (OTDR), 167, 208, 219-220
Bandgap, physical nature described, 114
Birefringence, stress induced, 167
Branch Exchanges (BX's), 216-217
Bulk Modulators, 224-225
Bowtie fibers, 171-172
Boundary conditions
  for circular dielectric cylinder, 99-100
  for rectangular dielectric waveguide, 98
Boundary value problem of a parabolic
    index fiber, 156-157

Carrier confinement, 123-124
Carrier lifetime in a semiconductor, 122
Chirp in semiconductor laser, 133-134
Cladding modes, 22
Classical dipole moment
  classical definition, 51
  as an expectation value, 61
Classical dynamics applied to laser dy-
    namics, 80-82
Complex dipole moment coefficient, 61
Complex index of refraction, 32-33
Conduction current, 31-32
Conductivity of a metal derived, 32
Confluent hypergeometric equation, 156

Constitutive relations, 13, 49
Coupled amplitude equations
  full form, 59
  regular form, 60
Coupled field matter equations, macro-
    scopic form, 64-65
Coupled mode equations, derivation
    from Maxwell's equation
  examples of solutions, 233-234
  finding solutions, 242-243
  for bent coupled section, 236-238
  for straight coupled section, 231-232
  solutions for simple $r_{51}$ region, 239
  solution for straight coupled section,
    232-233
  various forms possible, 243
Coupled polarization state equations
  derivation, 168-170
  solution and limiting cases, 170-171
Coupling coefficient for polarization
    states, 170
Coupling constant, $g_{\mu\lambda}$, defined, 69
Critical angle, 10
Current injection in a semiconductor,
    123-126

Dielectric absorption model, 34-36
Dielectric cylinder problem
  boundary conditions, 147
  comparison of hybrid mode propaga-
    tion constant with LP mode prop-
    agation constant, 161-162
  described, 98-99
  dispersion relation for propagation
    constant, 147
  dispersion relation for transverse elec-
    tric modes, 147-148
  dispersion relation for transverse mag-
    netic modes, 148
  expressions for transverse fields, 150-
    151
  impedance relation between longitu-
    dinal components of modes, 148-
    149

249

Dielectric cylinder problem *(continued)*
  longitudinal fields, 150
  relative magnitudes of the propagation
    constant, 149–150
  transverse mode structures derived,
    151–153
  weakly guiding dispersion relation,
    148
  weakly guiding dispersion relation for
    hybrid modes, 149
Dielectric tensor defined, 223
Dip in center of fiber core, 167–168
Dipole moment in a semiconductor, 121
Directed energy systems, 1–3
Directional coupler
  discussion of operating principles,
    230–234
  integrated optic version, 229–230
  transfer function, 243–244
Dispersion relation
  asymmetric slab, 38
  asymmetric slab waveguide, 26
  electron in periodic potential, 114
  free electron, 112
  slab waveguides, 23
  symmetric slab waveguide, 26
Dispersion relation graphical solution for
    symmetric slab, 26–28
Distributed loss approximation dis-
    cussed, 67
Double heterstructure, 122–124
Drawing of fiber, 167–168
Dynamical systems, 5–6

Effective conduction current, 37
Effective conductivity, 38
Effective index method
  application to general gradient index
    profiles, 107–108
  application to rectangular dielectric
    waveguide, 103–107
  applied to circular geometry, 139–140
  applied to semiconductor laser struc-
    tures, 136–137
  breakdown into one-dimensional
    problems, 104–105
  determination of field structure, 106
  determination of propagation con-
    stant, 104–106
  general theoretical exposition, 108–
    109

Effective index refraction defined, 18
Effective mass
  defined, 115
  derived from dispersion diagram,
    115–116
Effective mode index defined, 30
Effective temperature of a laser, 142–143
Electromagnetic energy density, 57
Electron wave functions above and
    below energy gap, 116–117
Electron free space wave function, 112
Electrooptic tensor defined, 223
Energy bands, reduced representation,
    117–118
Energy gaps discussed, 114–115
Equation of motion
  electron in a metal, 31–32
  of a laser state in a complex potential, 82
Evanescent field coupler described, 45
Expectation value of an operator, 49

Fabry–Perot resonator field expansion
    derived, 65–67
Fermi Dirac distribution, 118–119
Fermi level defined, 118–119
Fermi statistics defined, 117–118
Fiber optic gyroscope, 220
Free space optics, 1–3
Fresnel relations, 9–10

Gain function defined, 74
Geometrical optics
  applied to source coupling, 12
  discussion, 184
  in a slab waveguide, 8–12
Goos–Hänchen shift, 11
Group delay for alpha index profile,
    199–200
Growth of laser structures, 134–135

Hamiltonian, 52
Harmonic generation discussed, 222–223
Harmonic oscillator model for an elec-
    tron, 34–35
Hybrid modes (*see also* dielectric cylin-
    der problem)
  defined, 16
  relation to LP modes, 160–165
Hydrogen atom
  Bohr radius, 53–55

ground state energy, 53–55
Hydroxyl ion, 177

Imaging systems, 1–2
Incoherent excitation of a single mode
    fiber discussed, 193–194
Index of refraction in terms of dipole
    density, 36
Index tensor
    discussion, 168
    for $r_{51}$ electrooptic perturbation, 238
Indifussion, 227
Indirect gap semiconductors, why they
    will not lase, 127
Individual guide modes compared to
    system modes, 234–236
Integrated optic fabrication, 225–227
Interaction energy of dipole electromag-
    netic field interaction, 57
Interaction of matter and radiation, re-
    lated to bandgap, 120–121
Inversion condition in terms of quasi-
    Fermi level derived, 126–127
Inversion parameters
    defined, 61
    equation of motion for, 62
    in a semidconductor, 122
    relation between macroscopic and mi-
        croscopic forms, 64

Laguerre polynomials, defining equa-
    tion, 159
Lambertian source specific intensity
    defined, 193
Langevin terms, 78–79, 92
Laser dynamics in terms of two-di-
    mensional potential, 82–83
Laser modulation, discussion of effects
    associated with return to zero di-
    rect modulation, 219–220
Laser speckle described, 180–181
Leaky modes, 197–198
Lens waveguides, 1
Linearly polarized LP modes (see also
    optical fiber, LP modes), relation
    to hybrid modes, 160–165
Linewidths of fiber sources discussion,
    183–184
Longitudinal field component equa-
    tions, 96

Lorentzian line shape derived from
    rate equation parameters, 74

Mach–Zender interferometer
    discussion of operation, 229
    effects of perturbations on fabrication,
        242
    integrated optic version, 227
Material dispersion, 28, 201
Matrix method
    for bound modes of asymmetric slab,
        23–24
    for multilayer bound modes, 25–26
Maxwell's equations, 12–13, 48
Mesa etching discussed, 136
Modal expansion of a single waveguide
    mode, 155
Modal field expansion of monochro-
    matic and time-varying fields, 95
Modal noise
    calculation of statistics, 209
    discussion, 178–183
    phase and intensity statistics, 179–180
Mode definition, 14–15
Mode condition in WKB, 198
Mode continuum approximation dis-
    cussed, 188–189
Mode coordinate defined, 42
Mode expansion
    in terms of symmetric and antisym-
        metric modes, 230
    polarization state in a multimode
        fiber, 178
    polarization state of a single mode, 168
Mode parameter defined, 186–187
Mode spacing for parabolic index fiber,
    188
Mode sum
    for wavefunctions, 55–56
    general form, 29
Modes
    anti-symmetric, 228
    symmetric, 228
Modified chemical vapor deposition
    (MCVD), 165–167
Multiplexing/demultiplexing, 216

N-level systems, 87, 90
Normalization of modal power, 24, 29–
    31
Numerical aperture defined, 174

Optical fiber, (*see also* dielectric cylinder problem)
  comparison of LP propagation constants with hybrid mode propagation constants, 161–162
  coupling efficiency defined, 192
  cut-off condition in terms of principal mode number, 173
  discussion of actual polarization states, 154, 164–165
  discussion of stochastic nature of phases in a multimode fiber, 179
  excitation by incoherent source, 193–194
  excitation problem, 190–194
  expression for propagation constants of LP modes, 158, 172
  expressions for field structures of LP modes, 158–160
  linearly polarized (LP) modes derived, 154–161
  modal condition for linearly polarized (LP) modes, 158
  number of bound modes, 173
  numerical aperture defined, 174–175
  principal mode number defined, 173
  V number defined, 159
  relation of hybrid modes, 160–165
Optical path length changes induced by temperature variation, 182–183
Optical fiber loss, 176–177
Optical time domain reflectometry (OTDR) 167, 208, 219–220
Orthogonality relations for wavefunctions, 55
Oscillator strength defined, 35
Overlap factor $\Gamma$, 227
Overlap integral approximation discussed, 190–192

Parabolic index profile defined in terms of alpha index profile, 156
Phase shifter integrated optics, 225–226
Phenomenological damping discussed, 62–64
Photon lifetime
  in a Fabry–Perot cavity derived, 110–111
  related to loss parameter $\kappa\lambda$, 110–111
Photon normalization, 67, 70

Polarization evolution, 205–207
Polarization relaxation constant in a semiconductor discussed, 121–122
Polarization rotator
  composite device with birefringence matching, 240–241
  composite integrated optic device, 239–240
  integrated optic version, 238
Polarization vector
  defined, 49
  defined in terms of microscopic dipole moments, 64
Polarization for a semiconductor sketched, 133
Polarization in semiconductors, 141
Polarizability defined, 49
Polarization of a semiconductor compared with that of a gas, 131–133
Potential for a laser oscillator, 80–81
Potential function
  isotropic crystal lattice, 221
  non-centrosymmetric crystal lattice, 222
Poynting vector defined, 29
Power, electromagnetic defined, 30
Preform for fiber fabrication, 165–166
Principal mode number for alpha index profile, 199
Propagation
  in gain media, 41
  in a lossy dielectric, 36
  in a metal, 33–34
  in a semiconductor, 36–38
Propagation constant
  for alpha profile in WKB limit, 198
  in terms of ray coordinates, 186
Propagation phase in an electrooptic crystal, 224
Propagation vector in terms of ray coordinates, 185
Pumping of multilevel system discussed, 63

Quantum classical correspondence, 51
Quasi–Fermi level defined, 125

Radiation condition at infinity discussed, 97–98

Radiometry and specific intensity, 12
Rate equations
    derived from semiclassical equations,
        71–73
    general form, 73
    for single mode laser, 74
Ray equations
    in cartesian coordinates, 185
    in cylindrical coordinates, 186
Ray path quadrature for radially varying
        guiding index, 188
Rayleigh backscattered light, 167
Rayleigh scattering, 177
Rectangular dielectric waveguides
    approximate solution, 101–103
    comparison of approximate analytic
        and exact numerical solutions, 103
    derivation and discussion, 95, 96,
        101–107
Rectangular waveguide problem, separa-
        tion of variables, 97
Regional branch exchanges (RBX's),
        possible realizations, 217–218
Relations between transverse field com-
        ponents and longitudinal field
        components, 139
Relaxation oscillations discussed, 77
Repeaters, 217, 218
Rotating wave approximation, 68–69
Rotation matrix, 239

Scalar wave equation in polar coordi-
        nates, 155
Schroedinger equation, 51, 57
Sellmeier's relation, 201
Semiclassical laser equations
    equations, stability, 74–76
    first stated and discussed, 70–71
    in a semiconductor stated, 126
    restated, 94
    with spontaneous emission, 78
Semiconductor laser rate equations, 131
Semiconductor lasers, 6
Separation of variables applied to rectan-
        gular waveguide problem, 97
Slab waveguides, 4–5
Specific intensity defined, 12
Spontaneous emission
    effects thereof, 78
    rate in a semiconductor, 121

Stimulated absorption rate in a semicon-
        ductor, 121
Stimulation emission rate in a semicon-
        ductor, 121
System modes compared to individual
        guide modes, 234–236
Stationary states, 53
State probabilities, 56, 58

Telephone network discussion, 216
Thermal source dynamics discussion,
        79–83
Threshold, in terms of potential func-
        tion, 81–82
Threshold condition stated, 74, 76, 77
Total internal reflection, 4
Transverse confinement in semiconduc-
        tor laser structures, 136–137
Transverse electric (TE)
    boundary conditions, 19
    defined, 9
    polarization equations, 17, 19
Transverse field components
    in terms of cylindrical coordinates, 99
    in terms of longitudinal field compo-
        nents, 96
Transverse magnetic (TM)
    boundary conditions, 19
    defined, 9
    polarization equations, 17–19
Two-level system, 55, 58–59

V-number defined for symmetric slab, 27
Van der Pol equation, 79, 88
Variational principle
    for change in propagation constant
        due to electrooptic effect, 227
    for imaginary part of the propagation
        constant, 40–41
    for propagation constant, 227
Vector wave equation, 50

W fiber, 202–203
WKB approximation for transverse vari-
        ation of wave function in cylin-
        drical coordinates, 196–198
Wave function
    antisymmetric, 53, 56–57
    defined by Shroedinger equation, 51
    discussed, 49
    symmetric, 53, 56–57

Wave packet
  distortion due to dispersion, 213
  expression for lowest order dispersion
    in a multimode medium, 196
  expression for propagated version, 196
  for one-dimensional propagation, 194

Wave propagation in a lossy asymmetric
    slab discussed, 39–40
Waveguide dispersion
  defined, 28
  discussed, 201–202
Weakly guiding expansion discussed, 155

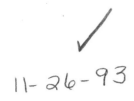